SpringerBriefs in Environmental Science

SpringerBriefs in Environmental Science present concise summaries of cutting-edge research and practical applications across a wide spectrum of environmental fields, with fast turnaround time to publication. Featuring compact volumes of 50 to 125 pages, the series covers a range of content from professional to academic. Monographs of new material are considered for the SpringerBriefs in Environmental Science series.

Typical topics might include: a timely report of state-of-the-art analytical techniques, a bridge between new research results, as published in journal articles and a contextual literature review, a snapshot of a hot or emerging topic, an in-depth case study or technical example, a presentation of core concepts that students must understand in order to make independent contributions, best practices or protocols to be followed, a series of short case studies/debates highlighting a specific angle.

SpringerBriefs in Environmental Science allow authors to present their ideas and readers to absorb them with minimal time investment. Both solicited and unsolicited manuscripts are considered for publication.

More information about this series at http://www.springer.com/series/8868

Julian Sagebiel · Christian Kimmich
Malte Müller · Markus Hanisch · Vivek Gilani

Enhancing Energy Efficiency in Irrigation

A Socio-Technical Approach in South India

With a Foreword by Prof. Dr. R.C. Agrawal

 Springer

Julian Sagebiel
Institute for Ecological Economy Research
Berlin
Germany

Christian Kimmich
Swiss Federal Institute for Forest,
    Snow and Landscape Research
Birmensdorf
Switzerland

Malte Müller
Institute of Agricultural
    and Horticultural Sciences
Humboldt University of Berlin
Berlin
Germany

Markus Hanisch
Institute of Agricultural
    and Horticultural Sciences
Humboldt University of Berlin
Berlin
Germany

Vivek Gilani
cBalance Pvt. Ltd.
Pune, Maharashtra
India

Additional material to this book can be downloaded from http://extras.springer.com.

ISSN 2191-5547   ISSN 2191-5555   (electronic)
SpringerBriefs in Environmental Science
ISBN 978-3-319-22514-2   ISBN 978-3-319-22515-9   (eBook)
DOI 10.1007/978-3-319-22515-9

Library of Congress Control Number: 2015947105

Springer Cham Heidelberg New York Dordrecht London

Printed on acid-free paper

Springer International Publishing AG Switzerland is part of Springer Science+Business Media
(www.springer.com)

Foreword

This work deals with the very timely theme of enhancing energy efficiency in irrigation, exemplified by a pilot project in the state of Andhra Pradesh in India.

Notwithstanding its declining contribution to the national gross domestic product, a natural corollary to the development process, the agricultural sector in India is still crucial to the all-round development of the nation. The sector currently employs nearly half of the population and has a critical role to play in the attainment of the national goals of increasing food security and reducing rural poverty. The temporal growth pattern of the Indian economy in the last decades bears out the direct and significant relationship to the state of agriculture today.

In the last fifty years, Indian agriculture has made tremendous progress, initiated by what is commonly known as the Green Revolution. Food production rose from 82 million tons in 1960–1961 to an estimated 263.2 million tons in 2013–2014. The Green Revolution was primarily characterized by employment of a package of practices—seeds, fertilizer, irrigation, and plant protection measures—to be supported by strong institutions. Irrigation occupied a pivotal role among these mainsprings of production growth, enabling the cultivation of two or more crops per year from the same piece of land. Due to huge investments in irrigation, the irrigated area in India now exceeds 63 million hectares, the largest of any country in the world.

However, the Indian irrigation system is highly inefficient. According to the Agricultural Outlook 2014–2023, jointly published by the United Nations Organization for Economic Co-operation and Development and the Food and Agriculture Organization (OECD-FAO), "India has one of the world's largest irrigation systems but it also faces high levels of inefficiency, particularly for those relying on surface water sources, the efficiency for which is estimated at 35–40 %, as opposed to ground sources, whose efficiency is estimated at 65–75 %. More serious is the problem of groundwater depletion, which is viewed to be in crisis as a result of excess extraction, due in part to the lack of regulated use and power subsidies which lower extraction costs".

The use of electrically powered irrigation pumpsets in India is increasing at a brisk pace of about half a million per year. More than 19.17 million pumpsets had been installed in India by the end of November 2014, with the figures for 1999 and 2009 being 12 million and 16 million, respectively. With increasing use of pumpsets, energy consumption for irrigation has also increased rapidly, growing at a compound rate of about 7 % between 2006 and 2012. India imports nearly a third of its total energy needs, with the government's Twelfth Plan estimating that it would need to import 29 % of its energy by 2016–2017, increasing to 31 % by 2021–2022, thereby putting heavy pressure on the national balance of payments. Oil subsidies put an additional burden—amounting to 0.8–1.1 % of the national gross domestic product in Fiscal Year 2013–2014—on the national exchequer.

Thanks to factors like abominable infrastructure, weak institutions, poor planning and implementation of projects, introduction of agricultural measures without adequately involving farmers, inappropriate equipment, and high subsidies, energy use in Indian agriculture is utterly suboptimal today. The average efficiency of pumpsets is estimated to be barely 30–35 %. However, through achievement of a stable electricity supply and more efficient pumping, the input of electricity for five-horsepower pumpsets could be reduced by up to 40 %.

The recent decline in global oil prices has somewhat eased the pressure on energy import costs for India, yet there is no room for complacency, and the necessity of enhancing efficiency in the use of energy and irrigation water is even greater, especially when climatic consequences are also taken into account.

This SpringerBrief seeks to make a valuable contribution in this direction through presenting the methods and results for a pilot project conducted in the Indian state of Andhra Pradesh. The design of the project is conspicuous by its incorporation and examination of the relationships between social, institutional, and technical variables. In observing that some social problems encountered during the project would not have occurred if certain technical problems had been absent and that these technical problems were able to be absorbed with proper social implementation, the necessity of intense and long-term relationships among various stakeholders for enhancing energy efficiency is highlighted. This reinforces the significance of one of the hitherto well-known but rather less-appreciated ingredients for the success of a development project: all stakeholders must be active participants throughout all of its phases and must also be made to feel involved in it.

Though the findings presented here relate to the state of Andhra Pradesh in India, the lessons have wider relevance. Farmers do not want cheap, subsidized, or free energy which is unreliable. They rather prefer to pay more for a timely, trustworthy, and stable energy supply. This would be a win–win situation for all stakeholders involved.

Berlin   Prof. Ramesh C. Agrawal

Preface

In 2008, the German Ministry of Education and Research launched the Future Megacities program, the aim of which was to identify scope for improvement in energy efficiency and climate change mitigation and adaptation strategies for rapidly growing megacities expected to reach a population size of ten million inhabitants within the next five years. Hyderabad, the capital of India's fifth largest state Andhra Pradesh,[1] was selected as one of these cities, with Humboldt-Universität zu Berlin, together with German and Indian partners, leading the project there. One focus, which became the theme of this SpringerBrief, was dedicated to challenges facing the power sector in Andhra Pradesh. As the agricultural electricity sector in Andhra Pradesh consumes about 30 % of total end-use in the state, it ends up playing a critical role for the urban electrical energy supply there. Consequently, the project consortium initiated a research agenda exploring possibilities for increasing energy efficiency in agriculture. Based on the findings from extensive field research, a pilot project was developed, the aim of which was, first, to understand existing agricultural electrical energy supply problems directly, from practice, and, second, to provide low-cost solutions which can be implemented independently of external funding. The relationship between social, institutional, and technical factors played a key role in the design of the project. Within the pilot project, about 800 shunt capacitors were installed to agricultural pumpsets used for irrigation in areas of rural Andhra Pradesh. Thirty farmer committees were formed, consisting of all farmers who participated in the project. The results were positive overall. Technically, an improvement of the power factor, an indicator of power supply quality, by about 16 % was measured, and field observations revealed an increased interest of farmers in the technology as well as regarding other aspects of irrigation and electricity. However, it was also realized that a narrowly technical approach can easily lead to failure, and intensive work with farmers is, in the end, a strong prerequisite for successful implementation. In practical

---

[1]On 2 June 2014, Andhra Pradesh was divided into two states, Andhra Pradesh and Telangana. As the pilot project ended in 2013, we will only consider the former state Andhra Pradesh in the SpringerBrief.

terms, severe problems with the capacitors occurred just after installation due to various reasons, including faulty maintenance and high-voltage fluctuations within the power system. This turn of events tested the robustness of the project in terms of social trust in the face of technical failures. It turned out that in villages, where the hold of the project was not strong, the project failed. Yet, in other villages, where more trust-building work had taken place, replacement of the failed equipment led to increased confidence among the farmers. All things considered, significant improvements can be achieved from upscaling the project. Assuming that all major electrically operated agricultural pumpset motors in Andhra Pradesh were to be equipped with a capacitor, overall energy savings could amount to 1,337 GWh per year, which would be equivalent to 1,216,623 tons of carbon dioxide equivalents emissions.

This SpringerBrief provides a comprehensive overview of the above-outlined project, including detailed description and analysis of how it was carried out. Background information on the power sector in India and Andhra Pradesh is also given, focussing on the special case of agricultural electricity supply and discussing strategies to improve it.

Project Background

The pilot project described here—Implementing Cooperative and Technical Solutions to Increase Energy Efficiency in Irrigation—was part of a research project on sustainable development in future megacities called Climate and Energy in a Complex Transition Process towards Sustainable Hyderabad: Mitigation and Adaptation Strategies by Changing Institutions, Governance Structures, Lifestyles and Consumption Patterns (hereafter, Sustainable Hyderabad). The Sustainable Hyderabad project was financed by the German Federal Ministry of Education and Research and consisted of the following German and Indian research institutions as its main partners: Humboldt-Universität zu Berlin; the Potsdam Institute for Climate Impact Research; Georg-August-Universität Göttingen; the nexus Institute for Cooperation Management and Interdisciplinary Research; and PTV Traffic Mobility Logistics AG, from the German side, and The Energy and Resources Institute, Delhi; Centre for Economic and Social Studies, Hyderabad; Osmania University, Hyderabad; International Crops Research Institute for the Semi-Arid-Tropics; and the National Institute of Technology, Warangal, from the Indian side. Additionally, each partner worked together with local bodies in Hyderabad, including ministries, governmental organizations, NGOs, other research institutes, and private consultants.

The Sustainable Hyderabad project's time frame ran between November 2008 and June 2013, focussed on different aspects of sustainable city development, including energy, water, transportation, food, health, and pollution. These topics were subgrouped into work packages and handled by the respective partners, each conducting their research from 2009 to 2011, including surveys, case studies, expert interviews, and theoretical calculations. The results of this initial work were used to initiate eight pilot projects from 2011 onwards, three of them in the energy sector. The Sustainable Hyderabad project came to an end in June 2013, issuing a Perspective Action Plan giving policy recommendations towards a more sustainable Hyderabad. A detailed description of the Sustainable Hyderabad project and additional information are available at www.sustainable-hyderabad.de.

Structure and Intention

This SpringerBrief outlines relevant aspects of the pilot project Implementing Cooperative and Technical Solutions to Increase Energy Efficiency in Irrigation in order to provide a basis for further discussion and implementation of such interventions. The overall aim of the project was to identify solutions for partly solving agricultural energy and water problems in Andhra Pradesh. The Sustainable Hyderabad project's research is focused on climate change adaptation and mitigation strategies, which the initiatives undertaken in the pilot project used as a primary guideline for implementation.

Here, the structure of this SpringerBrief will be summarized so as to guide readers on how best to read and understand it according to their interests. The SpringerBrief is divided into two main parts. *Part I: Background* deals with topics that are necessary for understanding the rationale of the pilot project, while also providing relevant information for readers who are not interested in the pilot project itself but want to acquire an understanding of topical issues in agricultural power supply, including solution strategies. *Part II: Pilot Project* presumes familiarity with the contents of Part I and explains the pilot project in detail. Readers who are already familiar with agricultural electrical energy supply in India, however, can start there directly.

Looking in more detail at the contents of this SpringerBrief, the first chapter introduces some basic concepts of power supply in India and briefly explains the persisting dilemma of low electrical energy quality for agriculture there. Chapter 2 provides information on the development of the power sector in particular Andhra Pradesh and India more generally, summarizing its current status with an emphasis on agricultural power supply and discussing the implications for farmers and other stakeholders of its flat-rate electricity tariff. Chapter 3 discusses strategies that can help reduce the power supply problem in this context. Section 3.1 summarizes recently completed and ongoing projects that have sought to improve the power supply for agricultural use in India. The Bureau for Energy Efficiency has, for example, initiated several large-scale projects which involve replacement of agricultural motors and initiation of high-voltage distribution systems. Apart from this, there have been smaller projects initiated by NGOs or universities trying to focus on farmers' involvement in managing power distribution. One example is the Lok Satta project, which established transformer committees for farmers in Andhra Pradesh. Section 3.2 discusses available options for improving farmers' supply situation, distinguishing between low- and high-cost solutions as well examining the interrelations between technical solutions and institutional requirements. We thereby draw a line between projects that aim to *replace* inefficient equipment, for example agricultural pumpsets, and projects that aim to *improve* the system with minor, but affordable technologies, even for farmers. Smaller solutions are more interlinked with the current institutional set-up than larger solutions and related technological changes, and a holistic approach demands the incorporation of technical and institutional solutions. Chapter 4 introduces some technical background information, explaining the Indian system of generation, transmission and distribution as well as the pumpsets, motors and capacitors in agricultural power supply there. This is important for gaining an understanding of some of the technical specifics that were part of the whole project's rationale. It is not necessary to be an electrical engineer to understand this chapter, as it is aimed to provide simple explanations reduced to the necessary facts and results. Readers who are aware of these basics can, however, skip the chapter.

Chapter 5 introduces Part II. Chapter 6 is perhaps the most important chapter in the entire document, as it gives an overview of all relevant topics required to understand the pilot project. First, the partners comprising the project team and the region where the project took place are introduced. Then, the stakeholders' aims, rationale, and technical and social approaches employed are explained and discussed. The technical and social approaches are discussed separately, though the project worked under the assumption that only a combination of both approaches could lead to project success. Chapter 7 summarizes the different steps in the project in chronological order, split into three phases: preparation and planning, implementation, and evaluation. The preparation and planning phase was used for undertaking intensive research in the project region in order to develop the overall concept and to select the technology, specific electrical feeders, and farming villages for the intervention. After having set up a detailed project plan, the implementation phase was initiated. This phase included awareness-raising meetings for farmers, installation of capacitors, and the establishing of farmer committees; we report on the conducting of this phase and discuss problems that arose during it. The evaluation phase primarily consisted of the measurement of technical parameters and was already initiated during the implementation phase. Different evaluation methods are compared, and the main hurdles encountered during evaluation are discussed here. The results of the evaluation are then discussed in Chap. 8, where we present the key performance indicators of the capacitors and use the resulting data for a marginal abatement cost analysis to compare the cost-effectiveness of the chosen solution in terms of carbon dioxide emissions with other available technologies, such as efficient motors and solar water pumpsets. Apart from the technical results, we briefly discuss some observations from the field, including what did and did not work. Finally, based on the results, Chap. 9 discusses the upscaling potential of the project, distinguishing between regional and technical upscaling and providing some ideas for a business model.

Finally, the last chapter summarizes the project and provides an outlook for further projects and research.

Work for the pilot project was complemented by several masters' and doctoral degree research investigations, some of the results of which have already been published in international journals and books. Throughout the text, the reader will find boxes summarizing some results of this research.

Relevance of this SpringerBrief to Other Areas and Contexts

In many countries, dependence on groundwater irrigation for agriculture is growing, while water and energy resources are becoming scarcer. Reasons for these tendencies are manifold and, in the context of climate change, irrigation is often considered as an adaptation measure, enabling farmers to be more independent of extreme heat waves, periods of no rain, and unpredictable weather events. But irrigation comes at the cost of increased usage of ground or canal water and energy resources, which are often not abundantly available either. Conditional on the institutional setting, energy in the form of diesel or electricity are the main inputs to power irrigation pumpsets. In Andhra Pradesh, one of the largest Indian states and the subject of this pilot project, groundwater irrigation is highly supported by local institutions, most obviously through the decade-old "free power to farmers" policy. As explained later in this SpringerBrief, such policies have created several dilemma situations or low-level equilibrium traps, where farmers, distribution companies, and the state as the cost bearer suffer from poor-quality electrical energy supply, high maintenance costs, and subsidy payments, respectively (Kimmich 2013). However, despite the very unique institutional situation, the problems farmers face in Andhra Pradesh are not very different to other states in India and many other agrarian countries. In particular in countries of the Global South, lack of financial capabilities, such as credits for suitable irrigation infrastructure, and social conflicts arising through the common pool resource characteristics of irrigation, similar problems as those in Andhra Pradesh, are observable. Researchers from various disciplines—including economics, the social sciences, and engineering—have conducted extensive research, providing a large range of possible solutions, including less resource-intensive technologies, incentive-based mechanisms, and collective action initiatives.

The concepts applied in the pilot project focussed on here have been adapted to the special conditions in Andhra Pradesh, yet many of its implications are generally valid. One main feature of the project was the formation of farmer committees to solve problems collectively. As the actions of farmers are interdependent, the behaviour of one farmer has effects on the outcomes of neighbouring ones. In our case, the unit of dependency was the distribution transformer, providing electrical energy to many farmers. Consequently, through inappropriate usage or over-pumping of water, one farmer can adversely affect the outcomes of others who are connected to the same transformer. Hence, we sought to find out whether managing groundwater pumping as a group could help towards overcoming such problems. During the pilot project, it became evident that farmers were able to collectively manage their distribution transformers and subsequent distribution systems in ways that are likely applicable to a variety of other contexts, even beyond agriculture and irrigation, as many kinds of development projects can be supported by collective action approaches. The key lessons learned from this project are, thus, not context-specific but rather valid everywhere where resources have public good characteristics. The pilot project itself relied on general results regarding collective action derived from various studies and experiments (see for example Ostrom 1990, 2005; Ostrom et al. 1994), thus benefitting from and then contributing towards further development of this field of inquiry.

The pilot project was focused on increasing energy efficiency. A simple technology, so-called shunt capacitors, was selected and installed into agricultural motors. The reasons for choosing capacitors instead of a broad range of other, perhaps more effective, solutions can be found in the specific conditions of agricultural power supply in Andhra Pradesh. In other areas, different technologies may suit the existing conditions better. Still, some important insights from using this particular technology may be valid for more general contexts, in that the project demonstrates the difficulties that can arise when introducing a new technology. Initial reluctance of stakeholders, lack of trust, and problems that arose due to technology failure are issues of a general nature, and the lessons learned from this project can be regarded as a guide to other projects aimed at working at the grassroots level on implementation of technological solutions.

The research community may also benefit from the pilot project's results. Although observations from applied projects sometimes lack scientific rigour, insights relevant to the behavioural sciences and the disciplinary interface between the natural and social sciences can be drawn from them. During the different phases of the project, complementary research was also being conducted, the results of which have provided insights regarding common behavioural patterns. For example, a framed field experiment was conducted with farmers, the aim of which was to better understand why cooperation sometimes fails, even if it promises better outcomes for all farmers. The research results from these investigations are currently being prepared for publication or are already published. This SpringerBrief provides an overview of the research conducted within the project and its main results.

To conclude, one thing has become obvious to those involved in the project: Projects aimed at enhancing development through new technologies need to seriously take into consideration the social dimensions of technological change and adaptation. This SpringerBrief seeks to demonstrate the validity of this assumption with reference to the pilot project's environment but with the intention of offering insights that may be relevant for many other contexts. The authors hope that readers can learn from the successes and failures of this project and use its findings to better design their own future projects.

References

Kimmich C (2013) Linking action situations: coordination, conflicts, and evolution in electricity provision for irrigation in Andhra Pradesh, India. Ecol Econ 90:150–158. doi:10.1016/j.ecolecon.2013.03.017

Ostrom E (1990) Governing the commons: the evolution of institutions for collective action. Cambridge University Press, Cambridge

Ostrom E (2005) Understanding institutional diversity. Princeton University Press, Princeton

Ostrom E, Gardner R, Walker J (1994) Rules, games, and common pool resources. The University of Michigan Press, Ann Arbor

Acknowledgments

This project would not have been possible without continuous support from local partners and stakeholders. We are more than grateful to Philip N. Kumar, who supported the project nearly from the beginning, visited the field regularly, talked with the farmers, resolved social conflicts, and made sure that the project would not stop at any point along the way. Likewise, we gratefully acknowledge the contributions of Vineet Goyal, Subash and Hari Krishna from Steinbeis, India, who made sure that the technical implementation ran smoothly. We especially thank Rama Mohan and Sreekumar, who advised the project in different phases and were always ready to listen to and comment on the project's progress. The technical realization in the field owes large thanks to Naveen, Tirupati, Illiah, and Ranjit, who installed, re-installed, de-installed, repaired, and maintained about 1200 capacitors. Bashkar, J. Mahesh, Maheshjee, Venkatesh, Kiran, and Nagraj guaranteed the social dimension of the project by forming 30 farmer groups. They never tired of going to the villages, talking with the farmers, organizing meetings, and making sure the team was always updated on the most recent developments. We further thank Krishna Reddy and Professor T.L. Sankar from the Administrative Staff College of India, who supported the project in Phase II with field visits and long discussions; Amit Jain and his team from the International Institute for Information Technology, Hyderabad, for their inputs; Dr. Ramesh Chennamaneni, who provided accommodation in Vemulavada; and the managing directors of CESS Sircilla, who supported the pilot capacitor installation.

We are grateful to Franziska Köhler, Kerstin Maas, and Marco Pompe, master's students who conducted their research within the project. Thanks also goes to Jens Rommel who critically commented on the project and frequently helped out in conducting research for it. We also thank Reinhold Wilhelm for assisting with the coordination from Germany, making sure that institutional hurdles were overcome.

We are greatly indebted to Dr. Amit Garg of IIM, Ahmadabad, for his guidance and continuous supervision, for lending his expertise throughout the duration of the project, and, crucially, his efforts at the completion stage. We would also like to extend our gratitude and thanks to industry experts Gyan Prakash and Nimit

Khungar, members of cBalance Solutions Pvt. Ltd., for their kind co-operation during the fieldwork, providing all the necessary support for project analysis and, last but not least, providing encouragement to all team members to complete this project. Special thanks goes to Marcus Mangeot and Casjen Ennen, who made a video documentary on the Sustainable Hyderabad project, for their great enthusiasm as well as for their ability to provide us with some diversion from the everyday life of the project. We are grateful to Christopher Hank for several hours of proof reading and Maximilian Kanig for creating a map.

Lastly, we would like to thank all farmers involved, who patiently listened in various meetings to the project team and enabled the realization of the entire project.

This pilot project has been conducted within the Sustainable Hyderabad project, financed under the Future Megacities programme of the German Federal Ministry of Education and Research (Grand Number: FKZ 01LG0506A1), from which we gratefully acknowledge generous financial support. We are also grateful to all readers who want to make use of our experiences, and we would be pleased to share project materials such as questionnaires and instructions when desired.

Contents

About the Authors

Julian Sagebiel is an economist specializing in international economics and doing research at the Division of Economics of Agricultural Cooperatives at Humboldt-Universität zu Berlin. Currently, he is conducting Ph.D. research on consumer preferences in the electricity sector, focusing on India and Germany. Since 2013, he has also been working as a researcher at the Institute for Ecological Economy Research in Berlin, where he focuses on valuation of ecosystem services and sustainable land management. julian.sagebiel@hu-berlin.de; julian.sagebiel@ioew.de

Christian Kimmich is an agricultural engineer and economist working as a postdoc at the Swiss Federal Institute for Forest, Snow and Landscape Research WSL. Previously, he had been a researcher at the Division of Resource Economics at Humboldt-Universität zu Berlin, where he worked on the regional governance of energetic biomass utilization, food versus fuel conflicts, and broader issues of ecological macroeconomics. He was a visiting scholar at the Ostrom Workshop in Political Theory and Policy Analysis at Indiana University and at the Center for Environmental Policy and Behavior at the University of California, Davis. christian.kimmich@wsl.ch

Malte Müller is an agricultural economist, Ph.D. candidate, and research assistant at the Division of Economics of Agricultural Cooperatives at Humboldt-Universität zu Berlin, Germany. His academic interest lies in the contribution of agricultural cooperatives to rural development and poverty reduction, as viewed within a broader context of development economics and cooperation. In collaboration with different research institutes, he investigates local phenomena related to cooperation and collective action in the field, with a focus on developing countries. malte.mueller@hu-berlin.de

Markus Hanisch is an agricultural economist with a background in institutional and resource economics. He is head of the Division of Economics of Agricultural Cooperatives at the Department of Agricultural Economics of Humboldt-Universität zu Berlin. His research interest combines approaches to political theory and institutional economics with cooperative studies. He is currently participating in or leading several research projects in close collaboration with national and international research foundations and financiers, such as the German Federal Ministry of Education and Research or the UN's Food and Agriculture Organization. He belongs to the group of Affiliated Faculty at the Vincent and Elinor Ostrom Workshops for Political Theory and Policy Analysis at Indiana University and works on several editorial boards. As a senior lecturer, he gives the course "Cooperation and Cooperative Organizations" within the EU-funded International Master of Science in Rural Development (IMRD) program. markus.hanisch@hu-berlin.de

Vivek Gilani is an Ashoka Fellow with an MS in Environmental Engineering from the University of Massachusetts. He has consulting expertise in the field of water and wastewater treatment design and analysis and has been certified as an energy auditor by the Indian Bureau of Energy Efficiency. In 2008, he co-founded the India-specific carbon footprint calculation and minimization body at no2co2. in. He is also the founder and director of cBalance Solutions Hub. Most recently, he co-founded The Green Signal—the first eco-labelling body in India—along with the Indian Institute for Management at Ahmedabad and the Center for Innovation Incubation and Entrepreneurship (CIIE). vivek@cbalance.in

All authors contributed to the pilot project presented in this SpringerBrief on improving agricultural electricity provision in India, within the German Ministry of Education and Research funded Emerging Megacity program. Julian Sagebiel coordinated the pilot project from 2011 to 2013. Christian Kimmich conducted his Ph.D. research on the sustainable provision of electricity for irrigation in agriculture from the perspective of evolutionary and institutional economics and game theory, providing the theoretical foundation for the pilot project. He accompanied the project throughout its duration. Malte Müller conducted the research for his master's thesis as part of the pilot project and coordinated the project between 2012 and 2013. Markus Hanisch was the initiator and head of the pilot project. Vivek Gilani was responsible for technical evaluation.

Part I
Background

Chapter 1
Introduction

Abstract This chapter introduces the contents of Part I of this SpringerBrief and highlights the vicious circle of agricultural power supply problems in India. The chapter starts with an introduction to the power sector in India, and Andhra Pradesh in particular, discussing its major challenges. Then, a brief overview of the agricultural power supply situation is given, followed by a short description of possible remedies to the currently existing low-equilibrium trap of low-quality power supply for irrigation.

Keywords India · Power sector · Agricultural power supply · Andhra Pradesh · Irrigation

A healthy power sector is often regarded as a key requirement for economic growth and foreign direct investment. Full electrification can act as a powerful tool for improving the livelihoods of the poor and a means to hinder rural–urban migration. India's power sector is one of the largest in the world and, over the last twenty years, has gone through major reform processes. In Andhra Pradesh, such reforms were initiated in 1991 as a response to a financial crisis at that time, during which the Andhra Pradesh State Electricity Board, the state-owned electricity provider, was running losses of about 1 % of the state's gross domestic product (GDP). In 2003, the Electricity Act was released by the Government of India, which provided guidelines for the way forward in the power sector, especially the promotion of renewable energies and a tariff system based on costs (Ministry of Law and Justice 2003; Ranganathan 2004). Additionally, the Ministry of Power established the Bureau of Energy Efficiency and respective state nodal agencies in 2002. In 2008, the Government of India released the National Action Plan for Climate Change, including the National Mission for Enhanced Energy Efficiency. Still, not all of the measures laid out there have materialised, and a large proportion of the country's consumers continue to face tremendous problems with their power supply. The least-resilient consumers are the rural population and farmers, with an electrical energy consumption of more than 30 % of the Andhra Pradesh

© The Author(s) 2016
J. Sagebiel et al., *Enhancing Energy Efficiency in Irrigation*,
SpringerBriefs in Environmental Science, DOI 10.1007/978-3-319-22515-9_1

total. The poor conditions of the transmission and distribution grid there frequently lead to high rates of motor burnout in agricultural pumpsets. Unbranded
and locally manufactured pumpsets, in combination with unqualified repairs,
decrease energy efficiency and further deteriorate overall power quality (Kimmich
2013).

It is widely understood that power supply for agriculture in India plays an
important role in current political debates. Agriculture is still considered to form
India's economic backbone, generating incomes for about 70 % of the population
and contributing to key political aims such as food security. Consequently, politi-
cians continuously promise farmers favourable policies to gain votes (Shah 2009).
Since 2004, farmers in Andhra Pradesh have received power on a flat-rate basis,
leading to a situation where incentives to invest in better equipment are distorted,
for both farmers and utilities, as farmers overuse the infrastructure and utilities
reduce their investments in it. This phenomenon can be described as a vicious circle of deteriorating power quality, leading to losses for utilities and reduced farm
output (Kimmich 2013). Taking this logic further, adverse effects with regard to
food security, groundwater overuse, and urban migration are becoming obvious.
Manifold strategies promoted by various stakeholders have been developed to
overcome this vicious circle, but the reality seems to remain unchanged.

Part I (from this chapter to Chap. 4) of this SpringerBrief outlines the main
concepts of the power sector in Andhra Pradesh and India, provides an overview
of its history and current status, and explains the situation of farmers in the context
of their increased dependence on groundwater for irrigation and, hence, their need
for a more reliable power supply.

In order to fully understand the situation of agricultural power supply in India,
and Andhra Pradesh in particular, it is important to examine the development of
the power sector since independence and the reasons behind the still-ongoing
reform processes. Until the early 1990s, the power sector was completely gov-
ernment-controlled. Each state operated through a State Electricity Board that
was responsible for generation, transmission and distribution. For several reasons, most State Electricity Boards became financially unhealthy already in the
1950s and were not capable of providing sufficient power in terms of either quality or quantity (Tongia 2007). Triggered by the Green Revolution in the 1960s,
electric groundwater pumping became popular (Shah 2009). Since then, the State
Electricity Boards have been increasingly burdened by excess power demand from
farmers and, as tariffs have not been cost-covering, unable to maintain sufficient
investment in infrastructure. As a consequence, power quality decreased over time,
which has led to the vicious circle described above. Even now, in most states in
India revenues from agricultural power supply are marginal or even negative, and
utilities are not capable of providing sufficient infrastructure. This historical development is explained in more detail in Chap. 2.

To understand why it has been so difficult to escape the vicious circle, one
needs to investigate previous attempts to break it. Most important have been government interventions. In 2006, the Bureau of Energy Efficiency defined standards

for pumpsets[1] and initiated several demand side management (DSM) programs, and state governments undertook efforts to improve the electric infrastructure in rural areas by, for example, introducing high voltage distribution systems, which reduce line losses and impede theft. Foreign donors like the United States Agency for International Development (USAID) started projects to train utility staff or to introduce new energy-efficient technologies (USAID 2011). Although many projects have achieved noticeable improvements, the overall goal of sufficient power for agriculture has not been attained. Neither, in many cases, no upscaling has taken place. Chapter 3 reviews these projects and then lists and discusses selected technical intervention options, including high voltage distribution system and small-scale technologies such as capacitors or energy-efficient pumpsets. It is important to distinguish between high-cost and low-cost interventions. High-cost interventions need to be initiated from above, meaning by the state government, and have to be implemented on larger scales. Meanwhile, low-cost interventions can be carried out on smaller scales, and farmers are able to participate in both their design and implementation. One advantage of the former is that no interaction with a local population is required and local conditions do not influence outcomes very much. However, there are interventions that can only be realised with farmer participation. Examples include learning the correct usage of pumpsets or implementing less water-intensive cropping patterns. The merits and demerits of high- and low-cost interventions are discussed in Sect. 3.2.

Finally, in order to understand the scope of the problem in India, one needs to grasp the interrelations between technical solutions and institutional requirements. Institutional approaches inherently require behavioural change. For example, training sessions with farmers can create greater awareness of water scarcity, which may, in turn, lead to more water preservation through adoption of other irrigation methods. In many cases, technical solutions only work when their institutional requirements are incorporated into the whole concept of change. The implications of this connection between institutions and technical solutions are discussed in the last part of Chap. 3. Chapter 4, meanwhile, complements the preceding chapters by explaining relevant technical details of the stages of the electricity process—from generation through distribution—as well the functioning of pumpsets, motors and capacitors.

References

Kimmich C (2013) Networks of coordination and conflict: governing electricity transactions for irrigation in South India. PhD Dissertation, Humboldt-Universität zu Berlin, Shaker, Aachen
Ministry of Law and Justice (2003) The Electricity Act 2003. New Delhi
Ranganathan V (2004) Electricity Act 2003—moving to a competitive environment. Economic and Political Weekly 2001–2005

---

[1]http://bee-dsm.in/PoliciesRegulations_1_4.aspx.

Shah T (2009) Taming the anarchy: groundwater governance in South Asia. Resources for the Future and International Water Management Institute, Washington and D.C. and Colombo and Sri Lanka

Tongia R (2007) The political economy of Indian power sector reforms. In: Victor DG, Heller TC (eds) The political economy of power sector reform. Cambridge University Press, Cambridge, pp 109–174

USAID (2011) Evaluation of DRUM and WENEXA, http://pdf.usaid.gov/pdf_docs/Pdacr528.pdf

Chapter 2
Background of the Agricultural Power Supply Situation in India and Andhra Pradesh

Abstract In this chapter, we discuss the power supply situation in India and Andhra Pradesh, beginning with a brief historical outline and then describing the current state and structure of the power sector, including its main challenges. We focus on agricultural power supply, exemplifying its major issues and discussing the existing low-equilibrium trap of power quality.

Keywords Power sector · South asia · Agricultural power supply · Irrigation · Low-equilibrium trap

2.1 History of the Indian Power Sector

Since independence in 1947, the power sector in India has been virtually controlled by the Government of India, which created State Electricity Boards that were responsible for the complete supply chain of power, including generation, transmission, and distribution. The reasons for this centralisation, based on socialist ideology, included no-monopoly instincts (profits were reinvested, fair-labour policy, no mark-up prices), economics of scale, control over price structure and the interconnection of State Electricity Boards to enhance system reliability (Tongia 2007). However, the State Electricity Boards turned out to be unprofitable and inefficient and, thus, required high subsidies from the Government of India and state governments to survive. The major reform process started in 1991 with a new government and an upcoming fiscal crisis. By then, the state deficit had reached 11 % of national GDP and, in order to maintain a growth rate of 8 %, high infrastructural investments were required, especially in the power sector.[1] It had become clear that there was hardly any scope for the Government of India to invest sufficient amounts by itself. Therefore, with help from the World Bank, it started to open the power sector to private and foreign investment. This, however,

---

[1] A general rule, which the Government of India was aware of, states that for a 1 % increase of economic growth a 1.5 % growth rate in the power sector is needed.

© The Author(s) 2016
J. Sagebiel et al., *Enhancing Energy Efficiency in Irrigation*,
SpringerBriefs in Environmental Science, DOI 10.1007/978-3-319-22515-9_2

did not mean the introduction of a competitive market. Rather, private investors faced restrictions but were guaranteed a 16 % rate of return, risk reduction and other benefits provided by the Government of India (Pani et al. 2007). Yet, many of the pursued investors stayed away at that time, and the projects that had been established often failed or led to even higher losses than the State Electricity Boards had before them. In the end, the private investment strategy turned out to be very expensive for the Government of India.

During the mid-1990s, the Government of India introduced further structural reforms (second stage of reform process), allowing the states to independently restructure their power sectors. State Electricity Regulation Commissions (SERCs) with a high degree of autonomy and responsibility (e.g., to set tariffs, resolve disputes, and monitor quality) were established, and the states started to unbundle their State Electricity Boards.[2] Andhra Pradesh, in the early 1990s unbundled with hardly any privatisation and is currently considered to be one of the leading states in terms of power generation and distribution (Sreekumar et al. 2007).

The third stage of the reform process was concerned with coordination and consolidation. The Government of India published the Electricity Act 2003 and established incentives for good performance, including ranking of states, competition among them, and rewards for the most efficient ones (Ministry of Law and Justice 2003; Ranganathan 2004). Another focus was directed towards the public with, for example, media campaigns like "power for all" being introduced. Additionally, the SERCs were asked to introduce full metering and to make sure their subsidies were paid back in time. Efficiency was also a target of the act. The Government of India had already established the Bureau of Energy Efficiency in 2002 and introduced new standards for efficiency. Further, private investors were encouraged to invest in a variety of sectors (Swain 2007).

But there has been strong opposition to such reforms, because power is regarded as a social good and many experts have feared that further privatisation would lead to higher electricity prices and limited access to energy for rural populations. In 2000, around 57 %, or 399 million, of the rural households and 12 % (84 million) of the urban households in India did not have access to electricity. By 2011, these numbers had decreased to 33 % for rural households and 6 % for urban households. In total, however, there were still 306 million Indians without access to electricity (World Energy Outlook 2013).

2.2 Structure of the Power Sector in India

Power is mainly generated by state-owned generation corporations (GENCOs) and few private companies. In Andhra Pradesh, for example, private generation contributes to 18 % of total production (Sreekumar et al. 2007).

---

[2]In this context, unbundling means that each stage of generation, transmission and distribution is carried out by a separate, independent company.

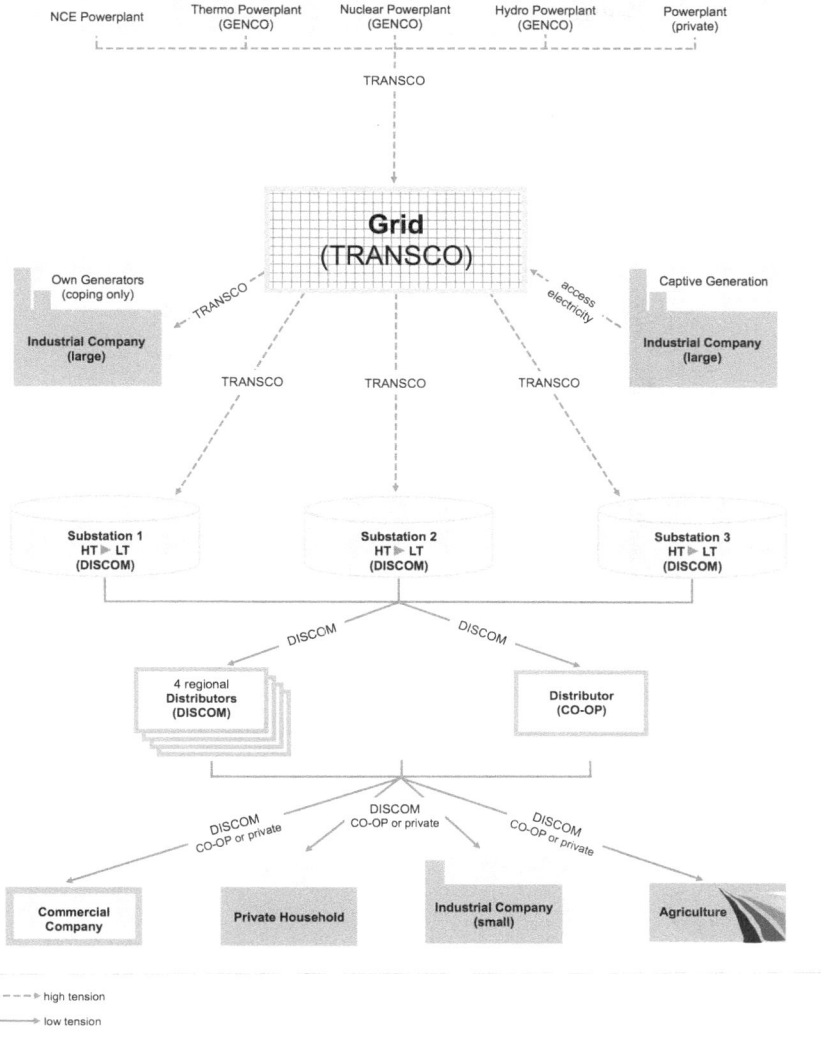

Fig. 2.1   The structure of the power sector in India

Transmission is provided by state-owned transmission corporations (TRANSCOs) on a high-tension basis to substations or directly to large electricity consumers such as the cement industry (Fig. 2.1).

At the substations, distribution companies, which are also state-owned in many states, take over, reduce the tension and distribute the electrical energy to distribution transformers (DTR), which finally forward it to consumers. In some states, private companies are allowed to conduct the final distribution. In Andhra Pradesh, this is not yet possible, but in some areas of Andhra Pradesh co-operative societies

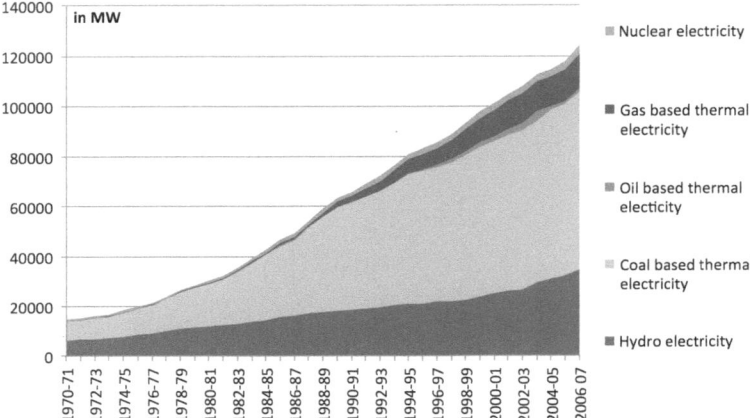

Fig. 2.2 Installed generation capacity excluding renewables in India by source, 1970–2007. *Source* Adapted from CMIE (2008)

that are responsible for final distribution and maintenance have been established (DRUM-Distribution Reform and Management 2006).

The major energy source that has been used up to now is coal (Fig. 2.2), which is abundant in India.[3]

Recently, a focus has been put on gas, as additional sources have been into view, and many private investors used gas, as generation facilities using it can be established very quickly. Renewable energy is being pushed by the Government of India (see Box 1) but still plays a minor role.

A large share of consumption comes from agricultural and industrial customers (Fig. 2.3). Noteworthy is that, despite investments into infrastructure, transmission and distribution losses account for up to 40 % of total generation, at least in some states.

The Government of India has paid roughly 250 billion Indian Rupees (INR) per year, that is about 1 % of GDP, for the losses of the now unbundled State Electricity Boards, with the direct subsidies alone adding up to 100 billion INR (Tongia 2007). Tariffs[4] are fixed and discriminate across consumers as a cross subsidy: private households and agricultural users pay less, sometimes nothing, and industrial and commercial users pay more. This is often regarded as a major source of end-use inefficiency. When industrial and commercial units face high tariffs, they tend to switch to captive power, which is more reliable but also more costly than power from the grid, leading to decreased competiveness (Ghosh and

---

[3]In many cases, domestic coal is of low quality—containing a high percentage of ash—and located in remote areas, which makes transport expensive, meaning that one needs more coal for "one unit of energy". This has led some companies to import coal.

[4]In India, a tariff refers to the price for electricity per kilowatt-hour, whereas in the United States electricity rates is the usual term.

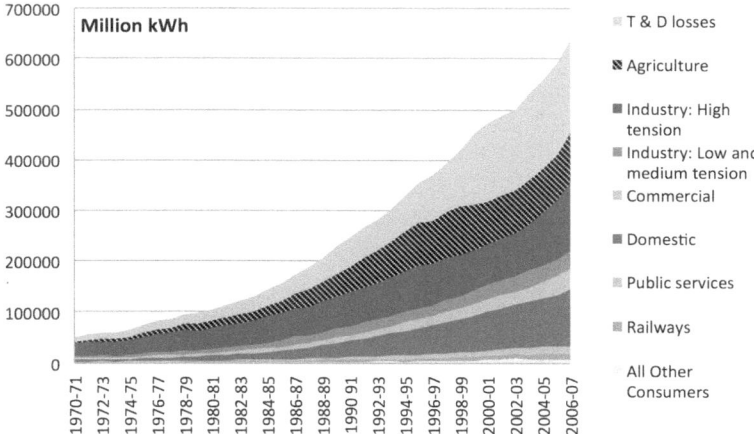

Fig. 2.3   Electric energy use in India by sector, 1970–2007. *Source* Adapted from CMIE (2008)

Kathuria 2014). Meanwhile, farmers, who often get power for free, use highly energy-inefficient assets for irrigation (TARU Leading Edge 2001). The next section discusses in detail the implications of providing free power supply to farmers.

First, however, we need to underline the main problem in the power sector: a continuous power-supply shortage. TRANSCOs and distribution companies are not able to supply at the normal voltage level (440 V for three-phase supply), which results in low-quality supply in the form of unscheduled power cuts, load shedding, fluctuating voltage and erratic frequency. Additionally, the low voltage levels lead to large technical losses and make power theft easier.

The problems of the Indian power sector can be summarised as follows:

- supply shortage, leading to power cuts and low-quality electricity;
- unsustainable and market-distorting cross subsidies;
- large-scale theft and non-payment of bills;
- inefficient and overstaffed utilities, suffering from a high degree of corruption;
- rural villages without access to energy services; and
- an incentive-distorting tariff system that cannot cover costs.

In the following chapters, the discussion will be reduced to the agricultural sector. However, it should become obvious how these problems interrelate with agricultural power supply. Solving agricultural power problems will immediately relieve the other sectors.

Box 1: Efforts to Expand Solar Energy in India

This box is adapted from Sagebiel et al. (2013)

In the Government of India's efforts towards resolving the power-supply problems, the Jawaharlal Nehru National Solar Mission (JNNSM) was initiated in 2010 under India's National Action Plan on Climate Change, aimed at increasing solar photovoltaic (PV) power generation in the country to 22 GW annually by 2022. As part of this process, rural areas are to be equipped with nearly 2 GW of off-grid installations. Under the Remote Village Electrification programme of the Ministry of New and Renewable Energy, 20 million square kilometres of land shall be used for solar PV collectors and 20 million solar lighting systems distributed to rural households. Apart from this, the JNNSM is geared towards facilitating research and development, increasing human capital in the field of PV and expansion of the solar-power manufacturing industry. One main strain on support, however, is financing. The installation of off-grid solar power is directly and indirectly subsidized through the National Bank for Agriculture and Rural Development (Sairam 2012). Off-grid solar systems with a capacity of less than 100 W peak and mini grids with less than 250 W peak receive a 30 % capital subsidy plus a subsidized loan. Harish and Raghavan (2011) criticise this approach, as it is discriminative in favour of smaller systems, whereas the relative costs of solar PV lighting systems decrease with increasing size. Gambhir et al. (2010) note that, although electrification of rural areas without grid connection is mentioned as a priority of the JNNSM, only 7 % of the subsidies were spent on off-grid solutions. Assessments of the first phase of the programme showed that the number of on-grid projects had increased significantly faster than rural off-grid projects. However, by 2012, about 500,000 small lighting systems and 700,000 solar lanterns had been distributed and 1100 MW of on-grid capacity had been installed.

Other governmental programs directly and indirectly facilitated the extension of solar PV power. Unbundling and privatization was one major prerequisite. The Electricity Act 2003 (Ministry of Law and Justice 2003) allowed decentralized power generation. Reforms in 2010 introduced renewable energy certificates and renewable power purchase obligations bind owners of transmission licenses to purchase 5 % of total power from renewable sources. Further, feed-in tariffs, which were introduced in most Indian states make investing in solar PV power attractive for the private sector.

2.3 The Vicious Circle of Agricultural Power Supply

As explained above, the reasons for the low quality of agricultural power supply can be found in the political economy of the Indian agricultural sector. Agriculture plays a crucial role in India's domestic economy, as about 70 % of the population generate their income from agricultural activities (Kimmich 2013a). The sector also fosters food security, and many industries, such as cotton, depend on inputs from it. As a consequence, Indian politics has put a great emphasis on agricultural development. Since the beginning of increased use of electricity for water pumping in the 1960s, the power sector has played a key role in agricultural policies. Farmers have demanded free power supply and, in many cases, received it, subsidised by the state governments. The rationale here has been that power supply is a fundamental requirement for modern irrigation, which is the main driver for increasing agricultural outputs. Kimmich (2010) identified three factors that have enabled the subsidising of agricultural power provision in Andhra Pradesh. First, the increased availability of tube-well technology for groundwater-based irrigation; second, the existing power infrastructure, the regulation of which has allowed political influence to be exerted by the incumbent party; and third, a form of inter-party competition that has led to a political contest for votes, with subsidies being a key campaigning issue.

The agricultural power subsidisation policy has enabled more secure food provision and also prevented food-price inflation. Yet, it has not only been economically inefficient overall but also triggered financial difficulties that led to a major change in the governance of power infrastructure in the 1990s. Possibilities for increasing groundwater availability and energy-efficient allocation for its pumping are inseparably linked within the power-irrigation nexus, and analysis of the political economy of the situation suggests that policy change can be most likely induced at the level of power distribution (Kimmich 2013a).

Taking these general conditions as given, the story of the vicious circle of low power quality can be explained more specifically as follows: flat rate power supply to agriculture has led to the use of inefficient pumpsets and excessive water pumping. In the majority of cases, capacitors or motor protection equipment are not being used, which further increases voltage fluctuations, resulting in a low power factor. Voltage fluctuations exist even at the substation level, and three-phase voltage is heavily imbalanced. The overuse of groundwater and power usually forces regulators to reduce power supply to off-peak hours. Often, power is supplied in two phases per day: one in the morning hours and one at night. The night phase has led farmers to use automatic starters. When current is switched on, most pumpsets thus start automatically and simultaneously, resulting in a heavy initial load that burdens the overall infrastructure.

Altogether, these dynamics have led to frequent motor and DTR burnouts and, in consequence, to increasing costs for farmers and utilities. In response, farmers have tended to use even less efficient, yet fluctuation-resistant, pumpsets, as financial incentives to implement DSM for improving energy efficiency are absent. Inefficient pumpsets reduce overall power quality, increasing pumpset and DTR damages, following which farmers and utilities face high repair costs (Tongia 2007). In fact, farmers often pay for DTR repairs themselves, even though they are owned by the utilities (Fig. 2.4).

Adoption of some DSMs—such as provision of standard-approved, ISI-marked, pumpsets with energy-efficient motors or capacitors—could reduce

Fig. 2.4 Farmers collectively repairing their DTR. *Source* Christian Kimmich

Table 2.1 Summary statistics for selected survey variables

Variable	Mean	Standard deviation	Median	Min	Max
Branded pumpset (1 = yes)	0.67			0	1
ISI-marked pumpset (1 = yes)	0.37			0	1
BEE-rated pumpset (1 = yes)	0.06			0	1
Capacitor successfully installed (1 = yes)	0.10			0	1
Motor burnouts per year	1.86	1.64	2	0	12
Costs for motor repair (INR)	2693	1513	2500	200	8500
Age of the pumpset (years)	7.21	5.94	5	0	30
DTR burnouts per year	1.02	1.04	0.70	0	7
Costs for DTR repair (INR)	621	870	400	0	8000

Source Adapted from Kimmich (2013b)

equipment damage and energy consumption. If implemented, such measures could help farmers and utilities to save on repairs and, due to increased energy efficiency, fiscal expenditures on subsidies could be reduced, contributing to the viability of agriculture and benefitting utilities as well as the overall economy through reduced fiscal burdens. Kimmich (2013b) has provided an overview of the share of adopted DSMs by farmers in four districts in Andhra Pradesh, including ISI-marked and BEE-rated pumpsets and capacitors (Table 2.1). Box 2 explains in detail why, although advantageous for all stakeholders, such DSMs have hardly been implemented in India thus far.

The political discussion on subsidised power for farmers is still ongoing and highly controversial. As Shah pointed out, "[t]he only link between the state and the millions of pump irrigators is electricity supply, over which the state has control" (Shah 2009, p. 142). It has also been suggested that institutional changes in regulation (Dubash and Rao 2008), together with physical innovations (Shah 2009) and pilot projects (Mohan and Sreekumar 2010), may enable efficient and equitable outcomes. The next chapter summarises some of the attempts that have been made and discusses different strategies regarding how to escape the vicious circle.

Box 2: The Core Action Situation: A Coordination Problem

This box is adapted from Kimmich (2013b, c)

Substations, covering several villages and distribution transformers (DTR), transform power to the 11 kV level. Depending on a DTR's capacity, between five and 25 pumpsets can be connected, each of which can negatively affect power quality. Exclusion of low-standard pumpsets is, however, difficult. Power quality, or lack thereof, spreads within the electric power distribution grid, affecting all users, as the decision of one farmer to use a low-quality pumpset affects all other pumpsets connected to the same DTR. Meanwhile, if all farmers choose to install low-quality pumpsets, the utilization of a standard-approved pumpset by only one farmer cannot improve power quality. Yet, if all farmers were to install a standard-approved pumpset, repair costs would be drastically reduced and all farmers better off. The use of a capacitor to balance out voltage fluctuations is subject to a similar coordination problem. Furthermore, if only one farmer uses a capacitor, equipment damages may often even *increase*, as "the equipment installed to increase [...] productivity is also often the equipment that suffers the most from common power disruptions. And the equipment is sometimes the source of additional power quality problems" (Dugan 2003, p. 2).

Unlike in a dilemma situation, however, no farmer will have an incentive to deviate from a better equilibrium, once reached, as standard-approved pumpsets and capacitors will tend to reduce equipment damages and improve pumping efficiency. A simplified bi-matrix model of the coordination problem at stake here highlights the two equilibria (i.e., Nash equilibria in pure strategies), marked with an asterisk. The equal payoff for the strategy not to invest ~I, and the loss incurred by the one not coordinating, makes this model type an assurance problem (Fig. 2.5).

Econometric analysis of this coordination problem reveals that, under the given conditions, the rational strategy is not to adopt any Demand Side

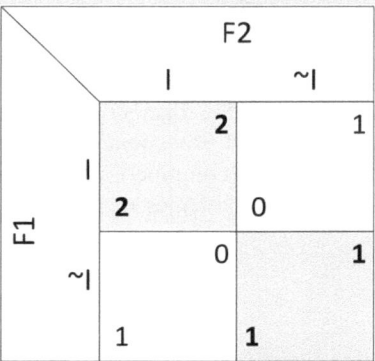

Fig. 2.5 Coordination problem of power quality measures as a 2 × 2 game matrix. Further description: Two farmers (F1; F2) have the choice to invest (I) or not to invest (~I) in measures to improve power quality. Outcomes are ordinal ranks. Investing (I) carries costs, but reduces equipment damage. If both F1 and F2 invest, the payoff is the highest, but if only one farmer invests, he carries the costs without gaining improvements in power quality

Measure solution. This is the low equilibrium of the underlying coordination problem, predicting no adoption at all. Yet, despite their negative impact on the frequency of equipment damage, a small share of the surveyed farmers has adopted Demand Side Measures, partly due to the legal order to make the use of them compulsory and related campaigns and partial enforcement when capacitors were distributed by the government.

Interviews conducted with farmers in Andhra Pradesh in 2010 indicate that only a few of them seem to understand how the Indian electricity system works as a whole. Thus, the interdependence of their decisions, especially in terms of their choosing to use non-standard pumpsets or not and the potentially positive outcomes that could result from simultaneous investment, do not appear to be conceivable for them.

References

CMIE (2008) Economic intelligence service—energy. Centre for Monitoring Indian Economy, Mumbai

DRUM-Distribution Reform U, Management (2006) Alternate participatory models for delivery in rural areas

Dubash NK, Rao DN (2008) Regulatory practice and politics: lessons from independent regulation in Indian electricity. Utilities Policy 16:321–331. doi:10.1016/j.jup.2007.11.008

Dugan RC (2003) Electrical power systems quality, 2nd edn. McGraw-Hill, New York

Gambhir A, Sant G, Deshmukh R (2010) Need to realign India's national solar mission. Economic and Political Weekly, Mumbai

Ghosh R, Kathuria V (2014) The transaction costs driving captive power generation: evidence from India. Energy Policy 75:179–188. doi:10.1016/j.enpol.2014.10.003

Harish SM, Raghavan SV (2011) Redesigning the national solar mission for rural India. Economic and Political Weekly, Mumbai

Kimmich C (2010) Policy persistence and the political economy of the electricity–irrigation conundrum in Indian agriculture: an analytic historical account. Paper presented at the colloquium series of the workshop in political theory and policy analysis, Indiana University, Bloomington

Kimmich C (2013a) Networks of coordination and conflict: governing electricity transactions for irrigation in South India. PhD dissertation, Humboldt-Universität zu Berlin, Shaker, Aachen

Kimmich C (2013b) Incentives for energy-efficient irrigation: empirical evidence of technology adoption in Andhra Pradesh, India. Energy Sustain Dev. doi:10.1016/j.esd.2013.02.004

Kimmich C (2013c) Linking action situations: coordination, conflicts, and evolution in electricity provision for irrigation in Andhra Pradesh, India. Ecol Econ 90:150–158. doi:10.1016/j.ecolecon.2013.03.017

Ministry of Law and Justice (2003) The Electricity Act 2003. Ministry of Law and Justice, New Delhi

Mohan R, Sreekumar N (2010) Improving efficiency of groundwater pumping for agriculture: thinking through together. Centre for World Solidarity, Prayas Energy Group, Hyderabad and Pune

Pani BS, Sreekumar N, Reddy MT (2007) Power sector reforms in Andhra Pradesh: their impact and policy gaps. Governance And Policy Spaces Project, Hyderabad

Ranganathan V (2004) Electricity act 2003—moving to a competitive environment. Econ Polit Wkly 2004:2001–2005

Sagebiel J, Kohler F, Rommel J, Kumar Goyal V (2013) Governance of solar photovoltaic off-grid technologies in rural Andhra Pradesh: some implications from the field. Energy and sun: sustainable energy solutions for future megacities. Jovis, Berlin, pp 27–36

Sairam R (2012) NABARD adds more power to solar mission. The Hindu, Chennai

Shah T (2009) Taming the anarchy: groundwater governance in South Asia. Resources for the Future and International Water Management Institute, Washington D.C., Colombo, Sri Lanka

Sreekumar N, Thimma Reddy M, Raghu K (2007) Strengths and challenges of Andhra Pradesh power sector: paper written for PMGER. Econ Polit Wkly 42:24–27

Swain AK (2007) Introducing competition in the Indian electricity: is micro-privatisation a possible way? Mimeo, New Delhi

TARU Leading Edge (2001) Andhra Pradesh power sector restructuring programme: baseline survey report. TARU Leading edge, Hyderabad

Tongia R (2007) The political economy of indian power sector reforms. In: Victor DG, Heller TC (eds) The political economy of power sector reform. Cambridge University Press, Cambridge, pp 109–174

World Energy Outlook (2013) World energy outlook electricity access database

Chapter 3
Strategies and Existing Projects

Abstract There have been various attempts, mostly put into practice through projects, to break the vicious circle of problems in Indian agricultural electrical energy supply. Such projects may be initiated by state governments, foreign development agencies, or are community driven. Projects with high budgets have often focused on large-scale technical interventions, where participation of local stakeholders is not required. In contrast, projects with low budgets have often involved local stakeholders and yielded low-cost technical solutions that can be implemented by farmers. In this chapter, we review recent and ongoing projects and discuss different implementation strategies.

Keywords India · Agricultural development · Irrigation technology · Agricultural projects · Research and development projects

3.1 Recent and Ongoing Projects in India

There are various approaches to and solution concepts regarding how to improve overall power quality in Indian agriculture. In the following, we review examples from four categories, including (1) public and state-level projects, usually driven by specific agencies, departments and ministries; (2) foreign-development cooperation projects; (3) research projects with a project-development component; and, finally, (4) community-driven approaches. Emphasis will be put on community-driven projects, as these follow an approach that is closest to the pilot project discussed in Part II of this SpringerBrief.

© The Author(s) 2016   19
J. Sagebiel et al., *Enhancing Energy Efficiency in Irrigation*,
SpringerBriefs in Environmental Science, DOI 10.1007/978-3-319-22515-9_3

3.1.1 Public and State-Level Projects

With its Agricultural Demand Side Management (Ag-DSM)[1] program, the Bureau of Energy Efficiency supports large-scale projects substituting old motors and pumpsets for new ones as well as investing in high voltage distribution systems. Further, Ag-DSM aims to find sustainable financial models, including private-public partnerships, for example through Energy Service Companies. In total Ag-DSM initiated 11 projects in different states covering a total of 20,000 pumpsets (Bureau of Energy Efficiency 2010). The first project took place in Solapur, in the State of Maharashtra, where about 2200 pumpsets were replaced with more efficient ones (Bureau of Energy Efficiency 2009). Based on the experience of these projects the Bureau of Energy Efficiency now expects that energy consumption can be reduced by nearly 40 % throughout India if all existing pumpsets are replaced by more efficient ones (Singh Saini 2013).

Projects are also being initiated by the state governments or agencies of the Government of India. In East Godavari District in Andhra Pradesh, the New and Renewable Energy Development Cooperation of Andhra Pradesh, a nodal agency of the Bureau of Energy Efficiency, initiated a project in 2009 for replacing DTRs with high voltage distribution systems. In total, nearly 200,000 pumpsets were covered (Planning Department 2013). In Gujarat, under the Jyoti Gram scheme, a project was launched to separate agricultural from rural-household feeders, a measure that was adopted as a precondition for stabilisation of rural supply (Shah et al. 2008). Also, over the last couple of years, the Ministry of Power has initiated several rural electrification programmes intended to positively affect agricultural power supply.

The Andhra Pradesh Electricity Regulatory Commission realised the importance of DSM in the late 1990s. The transmission and distribution companies committed themselves to distributing capacitors for agricultural pumpsets: "To improve the power factor, it must be made compulsory for the farmers to use capacitors with the pumpsets" (Andhra Pradesh Electricity Regulatory Commission 2000, p. 79), but they also realised that capacitors have been "the biggest techno-operational problem encountered in the power sector especially in Andhra Pradesh" (Andhra Pradesh Electricity Regulatory Commission 2006, p. 74). During a public hearing, a farmers' association was quoted saying that "though capacitors are purchased by the farmers, the licensees are reluctant to fix them, citing shortages in staff to do this work" (Andhra Pradesh Electricity Regulatory Commission 2010, p. 40). In recent surveys and discussions conducted by the authors, farmers continuously criticised the program for various reasons, including failure of capacitors, limited support from the distribution companies, and low-quality products.

---

[1]http://bee-dsm.in/.

3.1.2 Foreign Development Cooperation Projects

Foreign organisations, including the German Agency for Development Cooperation (Gesellschaft für Internationale Zusammenarbeit, GIZ), the German Development Bank (Kreditanstalt für Wiederaufbau, KfW) and USAID, have initiated several approaches intending to improve power supply in India, including agricultural supply.

USAID launched one of the largest of such efforts, the Distribution Reform, Upgrades and Management (DRUM) project, where about 25,000 engineers, managers and technicians from distribution companies were trained in technical concepts, management and project development, with a timeframe from 2004 to 2011 (USAID 2005, 2011). To follow up on the DRUM project, USAID initiated the Water and Energy Nexus Activity (WENEXA) project. Here, it was recognised that agricultural electricity improvements cannot be fully achieved without taking a closer look at the water sector and the interactions between energy and water. Within WENEXA, a number of pilot projects were initiated, including the installation of capacitor banks, replacement of motors and pumpsets and introduction of high voltage distribution systems. Meanwhile, GIZ has initiated various rural electrification projects, including solar photovoltaic systems and biomass power plants. KfW, on the other hand, focuses on financing large-scale projects through Indian governmental institutions, especially the National Bank for Agricultural and Rural Development, and financially supports investments in high voltage distribution systems.[2]

3.1.3 Research and Development Projects

More recently, projects have been linked to and designed on the basis of research from universities or research institutes, seeking to fulfil the overall goal of generating knowledge which can be used in further projects so as to make them more effective. Formally, the pilot project can be placed into this category.

The International Water Management Institute (IWMI), located at the interface between research and development projects, as a member research institute of the Consultative Group on International Agricultural Research (CGIAR), is working towards a solution for providing fixed amounts of energy on a metered basis. Such pre-paid metering could still be bundled with subsidisation, thus facilitating acceptance among farmers, while creating incentives to save power. It is hoped that this approach could simultaneously tackle the problem of groundwater overexploitation.

---

[2]An overview of GIZ and KfW projects can be found at http://www.giz.de/en/worldwide/368.html and https://www.kfw.de/International-financing/, respectively.

The Columbia Water Center, associated with Columbia University in New York City, launched a project in Gujarat in 2011 to create incentives for farmers to save water and, hence, power (Narula et al. 2011; Columbia Water Center 2013). The idea had been to provide financial benefits to farmers who reduce water and energy consumption. Together with the local utility and the state government of Gujarat, about 800 farmers were interviewed and asked to participate in the program. The farmers were provided with a fixed energy meter for their respective pumpsets. Those who reduced their consumption compared to a baseline from the interviews received monetary benefits, depending on the amount of reduction. Additionally, farmers were able to take advantage of various water-saving approaches, including trainings for better irrigation techniques and crop diversification strategies.

3.1.4 Community-Driven Projects

Other approaches have also been tried out, intending to improve power supply from the consumer side. The following presents three of them.

Village Committees: The XIMB Projects
The Xavier Institute for Management of Bhubaneswar (XIMB) found a lack of concern on the part of distribution companies in rural areas, as electrification is considered to be unprofitable for them due to fixed tariffs from the regulator, leading to a high burden for rural consumers, including billing on a load-factor basis instead of metering, low-quality supply, continuous power cuts and weak service provision from the utility (Mohanty 2002). The distribution companies had regularly ignored demands for improvement. At the same time, the consumers had not been united and were suffering from low bargaining power, with a commonly observed reaction being non-payment of bills and power theft.

A preliminary study conducted by XIMB revealed an immense distrust between distribution companies and customers (Mohanty 2002). The major finding of the study was that consumer action was "urgently required" for reading meters, understanding their bills and detecting theft.

In order to improve the situation, XIMB acted as an external facilitator, setting up 100 pilot projects in Orissa, India. Here XIMB established village committees, which were responsible for reading meters, collection of bills, dealing with complaints from customers, making decisions about new connections and disconnections as well as handling installation agreements (Dash 2006). In each participating village, a contact person was elected and made responsible for communication with the utility. The committees were given formal status, and monthly meetings with the utility were held.

According to XIMB, most of the pilot projects proved to be successful. Power supply and metering improved, bills were paid regularly, theft was reduced and service delivery improved. However, when the supplier did not act cooperatively or members of a village committee had obviously vested interests, improvements were low. By now, XIMB has extended its program to 4900 villages.

Micro-Privatisation

Swain (2007) combined the XIMB approach with a concept of micro-privatisation, where consumers would cooperate with the distribution companies via franchising. The idea was to put a small private company between the large supplier and the consumer in order to improve communication and enhance trust and cooperation. Swain argued that this model has the "potential to solve major problems in the sector like accessibility, mismanagement, theft, loss, and lack of transparency and accountability while providing choice for the users" (Swain 2007, p. 5). The XIMB projects often faced problems due to the unwillingness of the distribution companies to cooperate and, due to the high-costs of power supply and a subsidised tariff structure set by the regulator that did not cover costs, the incentives for higher-quality supply were low.

In Swain's scheme, the franchisee could serve as a hub and, being privatised, compete with other franchisees. Swain put an emphasis on competition and privatisation, with franchisees providing retail competition, which would then be expected to lead to higher-quality supply and cooperation with the local village committees. It is important within this approach that the franchisee acts under a distribution license rather than as an agent for the supplier. Swain also claimed that the user committees could form co-operatives, introduce capacity building and gradually take over a franchisee's license. He stated that the participation of users "can generate a collective preference for the service at the local level, reducing differences among individual users" (p. 21). This should then strengthen the bargaining power of consumers and, thus, might lead to improvements in quality and efficiency.

A Franchisee Approach: *Lok Satta*

Lok Satta is a political movement and party in Andhra Pradesh which has set up the Consumer Organisation for Regulation of Electricity (CORE), a part of the Consumer Advisory Committee of the Andhra Pradesh Electricity Regulatory Commission. The Central Power Distribution Co. Ltd., which is responsible for distribution in central Andhra Pradesh, permitted CORE to study three 33/11 kV substations in Andhra Pradesh. CORE found that maintenance was absent, technical standards were not being maintained, theft was rampant, meters were faulty and voltages were low (Rao 2006). Therefore Lok Satta, with assistance from the Administrative Staff College of India, set up four pilot projects, including private franchising. The objectives were to achieve

- improved quality of supply,
- metering of agricultural services,
- transparency and accountability,
- energy balancing,
- information gathering on agricultural consumption,
- alternative subsidy mechanisms, and
- replacement of inefficient pumpsets by energy-efficient ones.

Each project (East and West Godavari, Krishna and Guntur) covered the geographical area of an 11 kV feeder and all of the consumers served by it. A contract between Lok Satta (facilitator), the franchisee and the local utility (principal) was set up to give all authority to the franchisee. In many cases, the utility had already lost credibility with farmers. Consumer representatives held monthly meetings with all contract partners in order to understand each other's problems and preferences. The franchisee was responsible for taking care of breakdowns, preparation and collection of bills, monitoring and maintenance of meters, DTRs, all agricultural services and low-tension lines as well as education of farmers regarding efficiency and energy-saving approaches. Lok Satta recruited and monitored the franchisee, acted as a technical consultant and coordinated between the franchisee, the consumers and the local utility. The utility provided and repaired broken meters, lines and transformers based on the franchisee's monitoring, supplied the franchisee with data and guaranteed prompt payment to the franchisee. The project started in 2003, and a coordination committee and a monitoring commission were established. These had been meeting regularly and often solved existing problems successfully. Within three months, the consumers were satisfied with the franchisee, commercial losses had been reduced and revenue collection increased. Some consumers tried to manipulate the system "in the name of political influence", but this problem was often solved by the franchisee with Lok Satta support (Rao 2006). Nonetheless, the quality increases were not as high as expected.

3.1.5 Summary

The examples briefly described above are not exhaustive; yet they do give an impression of the current state of project designs and strategies. The approaches are rather heterogeneous and there have been many starting points. However, the majority of the foreign-aid projects work on a large budget, with solutions that usually are not implemented in a decentralised manner or on a local level. Ongoing foreign development projects emphasise technological innovations, while also taking into account educational factors. Meanwhile, public and state-level projects often have a focus on large-scale technical solutions.

In contrast, community-driven projects have been developed on a smaller scale, predominantly experimenting with institutional solutions. The three examples of such projects described here have emphasised the need for improvement in power supply and that consumer action could help to improve the situation. Although there were some difficulties from the consumer side as well as from the supplier side, experience has shown that, with aid from an external assistant, success is possible through consumer-participation approaches. Another important finding is that issues must be tackled simultaneously from the supply and the demand sides in order to be successful (Dash 2006).

3.2 Discussion of Various Implementation Strategies

In practice, a number of options are available for improving power quality and some of the main problems have been described in Sect. 2.3. Often, a solution to one problem creates a new problem. For example, more efficient water pumping may lead to an increase in groundwater usage, which in turn is likely to worsen the problem of scarce groundwater availability. Hence, each measure has to be examined carefully with respect to all of its foreseeable effects. In this section, some options are summarised and advantages and disadvantages elaborated upon. We distinguish between high-cost versus low-cost solutions and the complementarity of institutional and technical approaches. While many other dimensions could certainly be considered, we feel that our categorisation should provide sufficient insights for the purposes of this study.

3.2.1 Low-Cost Versus High-Cost Solutions

Implementation strategies can be classified into low-cost and high-cost technologies. While high-cost technologies usually require external financing and involve investments from utilities, low-cost technologies are generally affordable for farmers themselves. The current trend has been going towards high-cost solutions, like high voltage distribution systems, partly because they are assumed to solve overuse and power-quality problems at the same time. Low-cost solutions include, for example, power factor correction systems like capacitors[3] at the DTR level or directly at the load (i.e., motors) or mechanical devices that can help motors to run more smoothly, such as frictionless foot valves or improved pipes. One key characteristic of low-cost solutions on the demand side is that they require individual action by the user. Each farmer has to adopt a technology independently. Most high-cost solutions, in contrast, can be installed centrally, meaning that no direct involvement of farmers is required. One medium-cost technology, which also can be installed centrally, is the application of automatic power factor correction panels. This technology consists of several capacitors with different ratings, according to the size of the pumpsets connected. An electronic device measures the current power factor and switches on as many capacitors as are required. Hence, achievements in power factor correction are much higher than with the usage of single capacitors. The main disadvantage of the technology is its high maintenance costs. In field testing, the system has proven not to be stable and repair costs were very high. Additionally, as it turned out in discussions with manufacturers, an automatic power factor correction panel costs about ten times more than individual capacitors per kVA.

---

[3]A capacitor is an electrical circuit element that can correct the power factor in an electricity grid. It balances the phase between current and voltage (Dugan 2003; Meier 2006) and can, thus, improve power quality and energy efficiency (see Sect. 4.3).

Apart from solutions directly at the pump, energy can be saved via use of improved irrigation techniques, such as drip irrigation, or by using less water-intensive seeds. The most prominent example here is the system of rice intensification (SRI), which was introduced in the early 1980s (Surridge 2004). Using SRI, estimates of water savings, which are then directly linked to power savings, went up by 50 % compared to traditional methods of rice cultivation. Table 3.1 provides a brief description of some available technologies and their characteristics.

3.2.2 Technical Solutions and Institutional Requirements

Regardless of the technical solution, it is important to keep in mind that institutional (i.e., rules in place) and social factors play an important role. The users have to accept and, at least partially, understand the technology to make it successful. This includes technologies on the demand side where individual action is required, such as more efficient pumpsets and capacitors. But even with a centralised technology like high voltage distribution systems, farmers need to at least be made aware of it and should understand the key differences between it and their familiar system. Further, a high voltage distribution system does not substitute for power-quality measures, including demand-side capacitors and standardised, ISI-approved pumpsets. Hence, more efficient energy use and pumping should be combined with agricultural extension and training services, explaining for example the influence that pump use has on the power grid. This awareness can be combined with extension for optimal water management for each crop. Also, alternative crop patterns for best output from varieties of agricultural farmland and types of soil in relation to power use should be discussed.

A good example of an institutional innovation is prepaid electricity, where farmers can buy electricity in advance. This model requires, however, installation of meters at the pumpsets which, again, is not only a technical challenge but also requires acceptance and support from farmers. Given the political economy and the history of flat rate electricity provision, changing the institutional status quo is likely to be difficult.

Additionally, it may be of use to form farmer groups along the structure of the power grid that can, for example, manage water use and, together with local utilities, maintain "their" DTRs. Although the power infrastructure up to each DTR is managed and maintained by the respective distribution companies, collaboration is often crucial for reducing damages to infrastructure. In the ideal case, the farmers would collectively decide to invest in new demand-side technology and cooperate with their distribution companies. As each technological choice on the demand side has an influence on the grid, local distribution companies should, at the minimum, be informed when they are made. Because of the dangers of electricity shock, all installation and operation should be conducted by authorised staff only.

Table 3.1 Available technologies to increase energy efficiency in Indian agriculture

	Technology	Description	Main advantages/function	Disadvantages	Central/individual
Very high-cost	High voltage distribution system	Replace all DTRs and lines with High voltage distribution systems, so only three to five farmers are connected to one 25 kVA DTR	• Reduces line losses • Improves distribution • Reduces theft due to insolation • Reduces burn outs • Reduces maintenance work • Reduces poor tail-end voltages	• Requires large-scale intervention • More infrastructure is needed • Disadvantages might not be visible, as technology is new	Central
	Efficient pumpsets	ISI-approved (standardized) pumpsets and pumpsets with an energy efficiency of 50 % or more. Approved by Indian Bureau of Energy Efficiency	• Less electric energy consumption • More reliable water pumping • Less maintenance if operated correctly	• Does not work with high voltage fluctuations • Farmers have to rely on large companies for repair, rather than local repair shops • No option for changing rating manually	Individual
High-cost	Automatic power factor correction with DTR capacitor banks	Capacitor banks installed with a device that switches capacitors on and off as needed	• Maintains power factor of 95 to 99 % • Reduces line losses • Decreases reactive power • GSM technology allows for direct monitoring	• High maintenance costs • Vulnerable to breakdowns • Repair costs are high • Difficult to install	Central
	DTR capacitor banks	Power factor correction at the DTR level, with estimated reactive power required	• Improves power factor • Reduces line losses • Decreases reactive power	• As there are many DTRs, takes a long time to install • Not compatible with high voltage distribution systems	Central
	Capacitors at substation level	Power factor correction at sub-stations	• Improves power factor already at the substation	• High investment costs	Central

(continued)

Table 3.1 (continued)

	Technology	Description	Main advantages/function	Disadvantages	Central/individual
Low-cost	Capacitors at motor level	Installed to pumpset motor starters. Has an effect only if neighbouring farmers also use the technology	• Improves power factor • Reduces line losses • Decreases reactive power	• Maintenance by farmers is required • Can induce adverse effects (e.g., motor does not start)	Individual
	Frictionless foot valve	Installed directly to the pumpset	• Makes pumpsets more efficient • Does not require any electrical intervention	• Installation difficult	Individual
	Improved irrigation techniques	System of rice intensification (SRI) etc.	• No dependence on electric energy supply • Reduces energy demand • Can make farmers more competitive • Saves groundwater	• Requires intensive training of farmers • High initial adaptation costs • Poses high financial risks if technology fails • Incorporates uncertainties for farmers	Individual

These examples have been included here to show that technological changes cannot take place in an institutional vacuum and, often, not only imply behavioural changes and acceptance but also need to take into account legal requirements, including rights and duties. A theoretical discussion regarding this argument can be found, for example, in Bromley (1991).

References

Andhra Pradesh Electricity Regulatory Commission (ed) (2000) Tariff order: FY2000-01, Hyderabad
Andhra Pradesh Electricity Regulatory Commission (2006) Annual accounts 2002–2003, Hyderabad
Andhra Pradesh Electricity Regulatory Commission (2010) Tariff order: retail supply tariffs for FY2010-11, Hyderabad
Bromley DW (1991) Environment and economy: property rights and public policy. Basil Blackwell Ltd, Oxford
Bureau of Energy Efficiency (2009) Pilot agricultural demand side management (Ag-DSM) project at Solapur, Maharashtra
Bureau of Energy Efficiency (2010) Pilot agricultural demand side management (Ag-DSM) project at Muktsar and TaranTaran, Punjab
Columbia Water Center (2013) Groundwater depletion in Gujarat. http://water.columbia.edu/research-projects/india/gujarat-india/
Dash BC (2006) Governance of power sector: Orissa's experiments with village electricity committees. In: Economic and political weekly, pp 195–197
Dugan RC (2003) Electrical power systems quality, 2nd edn. McGraw-Hill, New York
Mohanty SB (2002) When customers become collaborators: an electricity distribution company's experiences in Indian villages. Xavier Institute of Management (India), Bhubaneswar
Narula Kapil, Fishman Ram, Modi Vijay, Polycarpou Lakis (2011) Addressing the water crisis in Gujarat, India. Columbia Water Center, Columbia
Planning Department (2013) Andhra Pradesh socio-economic survey report 201112. Government of Andhra Pradesh, Hyderabad
Rao PMM (2006) Rural power supply: micro experiments in Andhra Pradesh
Shah T, Bhatt S, Shah RK, Talati J (2008) Groundwater governance through electricity supply management: assessing an innovative intervention in Gujarat, western India. Agric Water Manag 95:1233–1242. doi:10.1016/j.agwat.2008.04.006
Singh Saini S (2013) Pumpset energy efficiency: agriculture demand side management program. Int J Agric Food Sci Technol 4:493–500
Surridge C (2004) Rice cultivation: feast or famine? Nature 428:360–361. doi:10.1038/428360a
Swain AK (2007) Introducing competition in the Indian electricity: is micro-privatisation a possible way?. Mimeo, New Delhi
USAID (2005) WENEXA: policy activities to reduce groundwater and energy use in irrigated agriculture in Andhra Pradesh, Washington
USAID (2011) Evaluation of DRUM and WENEXA, Washington
von Meier A (2006) Electric power systems: a conceptual introduction. IEEE Press: Wiley-Interscience, Hoboken, NJ

Chapter 4
Technical Background

Abstract In order to fully understand the main concepts and problems in Indian agricultural electric energy supply, some technical background knowledge is required. This chapter, written for readers without an engineering background, describes the concepts of generation, transmission and distribution and explains how agricultural pumpsets operate. Further, the concept of a power factor and the working of capacitors are also explained.

Keywords Transmission · Distribution · Electrical energy generation · Agricultural pumpsets · Capacitors · Power factor

4.1 Introduction

The evaluation phase of the project involved an extensive analysis of electric parameters to determine the measurable impact of the project on electrical energy consumption and quality. To lay a common ground for the technical discussions in Part II, this chapter seeks to provide basic insight into some of the technicalities involved.[1] To reach a wider readership, we aim to describe the processes without mathematical language and in a way that non-engineers should be capable of following. The more detailed electrical parameters and their mathematical interrelations are described in Appendix I.

To understand how greater efficiency in energy use in the agricultural sector of India is being achieved through irrigation-system improvements, requires a general understanding of the process of energy generation, transmission and distribution. As will be discussed in Part II, the project being described here specifically

---

[1]This chapter draws on the following, freely available, sources: Pacific Gas and Electric Company (1997), Beck and Martinot (2004), Bureau of Energy Efficiency (2005), von Meier (2006), Molburg et al. (2007), Michael (2008), Bureau of Energy Efficiency (2010), Bhatia (2012).

© The Author(s) 2016   31
J. Sagebiel et al., *Enhancing Energy Efficiency in Irrigation*,
SpringerBriefs in Environmental Science, DOI 10.1007/978-3-319-22515-9_4

aimed to assess the impact of capacitor additions at the pumpset level on parameters across electrical-energy distribution systems, especially between substations and pumpsets. This chapter highlights the role of capacitors in attempting to correct the systemic low power factor caused by inductive loads created by pumpset motors; capacitors can help in reducing the reactive power of the motors, thereby lowering the current drawn by pumpsets for performing a given task. How this happens will be explained in more detail below.

4.2 Electricity Infrastructure

In a power infrastructure system, the energy flow from source to load is composed of three stages: generation, transmission and distribution. These stages depend on each other to operate smoothly. Generation refers to the process of transforming different types of energy sources into electrical energy, whereas transmission and distribution mean the transportation of the electrical energy to the end user, or load. The main distinctions between transmission and distribution are the length of power lines and voltage levels. Transportation losses are lower the higher the voltage is. Therefore, voltage is stepped up after generation, transported through transmission lines to substations, stepped down and transported further via distribution lines to the end user. Usually, the generated electrical energy has low voltage, at 11/33 kV, which is then stepped up to 132/220 kV or higher for long-distance transmission to reduce losses. It is then stepped down to 33 kV at transmission substations and transferred to distribution substations. From there, it is stepped down again to 11 kV and, through distribution lines and DTRs, distributed at the 440 V level to the loads.

4.2.1 Generation

Electrical energy is generated from diverse energy sources, which include fossil fuels, such as oil, natural gas and coal, and renewable sources, like solar, wind and hydro. Thermal, hydro and nuclear power plants are the conventional methods used in India and usually generate power in bulk. Some of the alternative sources of energy employed there include wind, the sun, tidal waves or gas from biomass resources.

The total installed capacity of electrical energy generation in India was 266,644 MW in 2013.[2] Because of easy accessibility and a relatively low unit cost, coal and other fossil fuel-based thermal power plants contributed 179,072 MW,

---

[2]All figures presented in this chapter are taken from Ministry of Statistics and Programme Implementation (2014).

roughly 70 % of total generation capacity. Hydro power plants contributed about 15 %, with an installed capacity of 39,491 MW in 2013. Under the given conditions, there has still been high growth potential for hydro power plants in India. Yet, due to high investment costs and unavailability of transmission lines, many hydro projects are still in pipeline status. Contrary to many industrialised countries, however, nuclear energy contributes only 1.79 % to India's total generation.

In Andhra Pradesh, the installed electrical energy generation capacity through thermal power plants and hydro power plants is 8950 and 3730 MW, respectively. The high share of thermal power generation contributes significantly to India's CO_2 emissions. On the other hand, hydro and nuclear power plants have lower direct emissions but do pose safety and land use change problems.

4.2.2 Transmission

A functioning transmission system is a prerequisite for stable power supply. Electrical energy generated at power plants needs to be transmitted to load centres and a variety of consumers via transmission lines and towers. Power transmission lines are composed of two or more conductors to transmit electrical energy from one facility to another. Electrical energy is transmitted at higher voltages and lower current, since higher voltages reduce drops in line resistance and reactance. Generally, the longer the distance of the transmission lines the higher is the selected voltage.

Transmission lines are subject to line losses, meaning that during transmission electrical energy is transformed to other forms of energy and cannot be used at the end use appliance. There are three kinds of loss—dielectric, copper and radiation-induced—which are related to heat generation during transmission and surrounding electromagnetic fields. In India, these transmission and distribution losses are relatively high at about 23 % (see Fig. 2.2), compared to most developed countries at 2–6 %. The main causes for such high losses in India are old transmission lines, poor maintenance and line congestion, but theft is also covered under this statistic. Line congestion occurs when too much power is transmitted at one time, which leads to greater heating and resistance of conductors, thereby increasing line losses. Furthermore, voltage drops significantly under congestion and quality for end users deteriorates.

There, however, are ways to reduce line losses and congestion. If congestion is present on a permanent basis, an extension of transmission lines is required. But also real time monitoring and scheduled maintenance, better equipment at substations, including capacitor banks and more high voltage distribution systems, can effectively reduce losses. Further, increased use of decentralised renewable energy generation can help to relieve pressure on transmission lines, as production and consumption take place at the same location.

4.2.3 Distribution

The distribution grid starts at the distribution substation, where power is reduced to 11 kV before being sent out. It is then distributed to DTRs, which step down the voltage to the levels required for end users. For agricultural and domestic users, this is 440 V for three-phase supply and 230 V for single-phase supply, with agricultural pumpsets requiring 440 V. Substations often use separate feeders for industrial, domestic and agricultural supply. For agriculture, meters at a particular substation are the last means for measuring consumption, as motors are usually not metered (Fig. 4.1).

Hence, it is difficult to estimate actual agricultural consumption. Distribution lines in rural areas are often in poor shape, suffering from congestion and lack of maintenance, which is often taken care of by local populations. Low-hanging lines also pose health risks and their damaging can lead to supply interruptions.

Fig. 4.1 Agricultural DTR with meter in Andhra Pradesh, *Source* Malte Müller

4.3 Agricultural Pumpsets

Two types of pumpsets, centrifugal and submersible, are typically used for agricultural irrigation. Centrifugal pumpsets are used more frequently, because investment costs for them are lower than for submersible pumpsets, but they are usually operated in shallow wells (Fig. 4.2). A centrifugal pumpset is relatively simply constructed, consisting of an impeller and a diffuser. The impeller is the only moving part in the pumpset and is directly powered by the motor through a shaft, creating the (head and) pressure that is required to draw water. A diffusor then directs the water.

A submersible pumpset operates in a vertical position and consists of a pump bowl, a directly attached motor, a discharge column and a head assembly. It is used for bore wells and is installed submerged into them. As these are usually

Fig. 4.2 Shallow well near Hyderabad, *Source* Malte Müller

much deeper than shallow open wells, the motor is not easily accessible, so repairing and maintaining a submersible pumpset is more laborious than it is with a centrifugal one. Nonetheless, a submersible pumpset is more efficient, because the motor is directly attached, thus reducing friction, and cooling is provided by the water in which it is placed. It is also more resistant to changes in water levels and can be used in very deep wells. When used for irrigation, both types of pumpsets have a rating of between 2 and 15 kW.

The overall energy efficiency of a pumpset is influenced by the quality of the power input, the efficiency and friction of its motor, the efficiency of the pumpset itself, and the layout and design of the piping system. Efficient pumpsets have less copper and iron losses and reduce inefficiencies at the extreme ends (head and flow). The efficiency of a pumpset depends also on field characteristics. Efficiency losses occur, for example, when the size of a pumpset is not adequate for the conditions. In many cases, pumpsets are oversized and are, consequently, not being operated at the optimum load.

4.4   Power Factor and Capacitors

The power factor is defined as the ratio of active or real power, measured in kW, and apparent power, measured in kVA

$$Power\,factor = \frac{\text{kW}\,(Active\,Power)}{\text{kVA}\,(Apparent\,Power)}$$

Apparent power is also known as total power and refers to the power that is actually being provided to a system. Active power is the share of the apparent power that can be transferred into other, productive forms of energy. Unused power is called reactive power, which is minimised in properly working distribution systems.

A power factor of 1 (unity) indicates that all electrical energy is being used at a given point, meaning that no reactive power is present. Meanwhile, a power factor of 0.6 means that only 60 % of the apparent power can be used. In other words, to run a 3 kW motor, one would require $3/0.6 = 5$ kVA. In other words, a 100 kVA DTR is capable of powering twenty 3 kW motors when the power factor is 0.6 and 33 motors when the power factor is unity. A low power factor, thus, implies additional generation requirements for the same level of output.

Three-phase inductive loads required by larger electrical motors need the three phases to be balanced to reach a high power factor. The larger the imbalance is, the lower the power factor. Overloaded DTRs, wrongly sized motors, increased line voltages and already-existing imbalances at substations can lead to low power factors. In addition to being able to resolve these issues, capacitors can also help to correct the power factor.

A capacitor consists of two parallel metal plates with a gap in-between, which is filled with (dielectric) insulating material. This construction allows it to store and release electrical energy. The storage ability of a capacitor is called its capacitance. Static capacitors, as used in this project, have a fixed capacitance. Based on a given uncorrected power factor, one can calculate the required capacitance needed to reach a power factor of unity.

References

Beck F, Martinot E (2004) Renewable energy policies and barriers. In: Cleveland CJ (ed) Encyclopedia of energy. Elsevier, New York, pp 365–383

Bhatia BE (2012) PDHonline Course E144: power factor in Electrical Energy Management

Bureau of Energy Efficiency (2005) Guide books for energy managers book 3: energy efficiency in electrical utilities

Bureau of Energy Efficiency (2010) Pilot agricultural demand side management (Ag-DSM), Project at Muktsar and TaranTaran, Punjab

Michael AM (2008) Irrigation: theory and practice. Vikas Publishing House Pvt Ltd, New Delhi

Ministry of Statistics and Programme Implementation (2014) Energy statistics 2014. Government of India, New Delhi

Molburg JC, Kavicky JA, Picel KC (2007) The design, construction, and operation of long-distance high-voltage electricity transmission technologies. Argonne National Laboratory, Chicago

Pacific Gas and Electric Company (1997) Agricultural pumping efficiency improvements

von Meier A (2006) Electric power systems: a conceptual introduction. IEEE Press, Wiley, Hoboken

Part II
Pilot Project

Chapter 5
Introduction

Abstract The pilot project Implementing Cooperative and Technical Solutions to Increase Energy Efficiency in Irrigation was initiated in 2011 as a research and development approach toward exploring low-cost solutions for improving energy efficiency in Indian agriculture. It was developed under the assumption that technical solutions only work when accompanied by social interventions and capacity building. The technical solution consisted of installing shunt capacitors for agricultural pumpset motors where a low power factor had been observed. Capacitors are a low-cost solution which farmers can easily afford themselves, providing benefits to farmers and local utilities by reducing electrical energy consumption and motor and distribution transformer burnouts. This chapter introduces Part II of this SpringerBrief, outlining the pilot project steps and rationale.

Keywords India · Agricultural power supply · Pilot project · Capacitors · Social intervention · Capacity building

The pilot project discussed here was initiated in 2011 as a research and development approach to exploring low-cost solutions for improving energy efficiency in Indian agriculture. It was developed under the assumption that technical solutions only work when accompanied by social interventions and capacity building (see Sect. 3.2). The technical solution consisted of installing shunt capacitors at motors for agricultural pumpsets where a low power factor is observed. Capacitors are a low-cost solution which farmers can easily afford themselves, providing benefits to farmers and local utilities by reducing electrical energy consumption and motor and DTR burnouts.

The pilot project was divided into three phases: planning and preparation, implementation, and evaluation. In the planning and preparation phase, the project initiator—the Division of Cooperative Sciences at Humboldt-Universität zu Berlin—formed a project team consisting of engineers and social scientists. The project team then decided during two workshops and several discussions on the details of the intervention. In the implementation phase, the technical partners

© The Author(s) 2016   41
J. Sagebiel et al., *Enhancing Energy Efficiency in Irrigation*,
SpringerBriefs in Environmental Science, DOI 10.1007/978-3-319-22515-9_5

installed the capacitors and provided technical support to the farmers. Before and during the installation period, the social partners worked together with the farmers and formed farmer committees, which aimed at assuring farmer support for the project. Further, farmers were motivated to participate actively, and selected farmers were trained in technical issues related to their pumpsets, such as how to properly maintain a motor. The committees also aimed at strengthening the "voice" of the farmers, creating awareness and knowledge regarding power systems, as well as providing a basis for further interventions. During the implementation phase, the project team had already initiated the evaluation phase, which was mainly concerned with measuring the effects of the capacitors on selected indicators to get an idea of their impact on the energy efficiency, water flow and electrical energy consumption of the pumpset motors.

The project was, however, burdened by a technical problem. Two months after installation, the capacitors started to burn, sometimes destroying the whole starter box in which they were installed and, therefore, many farmers started to uninstall the capacitors. The project team decided to replace all installed capacitors with an improved version that consisted of a box with additional safety equipment. The problem led to an extension of the project in terms of budget and time.

Overall, the project team regards the results as being positive, as our evaluation has revealed improvements in technical parameters, and a subsequent marginal abatement cost analysis showed capacitors to be a very cost-efficient tool. Possible alternatives, such as efficient motors and solar water pumpsets, lead to significantly higher costs per kilogram of carbon dioxide (CO_2) abated and units of electrical energy saved. However, it should be noted that the effects of capacitors are limited. If India wants to reduce CO_2 emissions and electrical energy production to anticipated levels, only installing capacitors would not be sufficient.

The next chapter sketches an overview of the project, providing insights on the project's location, partners, rationale and aims. Chapter 7 details all steps that the pilot project has gone through, chronologically, while Chap. 8 explains the results, especially in terms of technical evaluation, and Chap. 9 discusses upscaling potentials.

Chapter 6
Project Overview

Abstract The chapter starts with a description of the project partners and region and then explains its aims and the roles of the stakeholders. The technical and social approaches adopted are also introduced, explaining why capacitors were selected for the technical intervention and how, in order to assure a smooth implementation process, farmers were organized into distribution transformer committees.

Keywords Vemulavada · Karimnagar · Socio-technical   interventions · Shunt capacitors · Distribution transformer committees

6.1 Partners

The multi-stakeholder project was constructed in such a way that several partners from different backgrounds were able to aggregate their knowledge to cover all of its relevant aspects. The consortium consisted of

- the Division of Cooperative Sciences at Humboldt-Universität zu Berlin (COOP), as the project leader and initiator;
- the Co-operative Electric Supply Society Sircilla, Ltd. (CESS), the local utility;
- the Self Employed Welfare Society, a local NGO, leading a watershed program in the area;
- the Power Systems Research Centre of the International Institute for Information Technology, Hyderabad (IIIT-H), the academic partner;
- the Steinbeis Centre for Technology and Innovation (SCTI), the technical implementer;
- the cBalance Solutions Hub Pvt. Ltd., the evaluator; and
- the Centre for World Solidarity (CWS) and the Prayas Pune Energy Group (PEG), the advising partners.

© The Author(s) 2016   43
J. Sagebiel et al., *Enhancing Energy Efficiency in Irrigation*,
SpringerBriefs in Environmental Science, DOI 10.1007/978-3-319-22515-9_6

Each partner had a specific and unique role in the project, as described below.

COOP

The Division of Cooperative Sciences (since 2014, Division of Economics of Agricultural Cooperatives) is part of the Faculty of Life Sciences at Humboldt-Universität zu Berlin. Its main area of expertise lies in the analysis of economic and social coordination and cooperation failure, which serves as a theoretical basis for applied research projects being carried out in several countries, including India, Uganda, Guatemala, and Kyrgyzstan. COOP led and coordinated the pilot project and was responsible for its overall implementation and evaluation, with its main activities being to

- coordinate all partners and activities;
- establish a structure to guide the carrying out of the project;
- document and evaluate the project;
- conduct a social survey; and
- establish farmer committees, including extensive training and capacity building events.

CESS

The Co-operative Electric Supply Society Sircilla, Ltd. was established in 1969 as part of an Indian governmental programme to enhance the spread and effectiveness of rural power distribution and management in India. One of the first five rural electricity supply companies in India, CESS now has an area of operation that extends over 173 villages, 109 hamlets and nine urban centres, currently serving more than 1.6 million customers altogether. The main activities of CESS in the project were to

- provide local knowledge to the project team;
- support installation of capacitors and measurement at DTRs; and
- participate in farmer meetings.

SEWS

The Self Employed Welfare Society is a non-profit organisation located in Vemulawada, in Karminagar district. It has launched many schemes, such as watershed programmes and clean drinking water initiatives. Its main project activities were to

- serve as an intermediary in the field to coordinate and communicate with farmers;
- report on and resolve problems/issues raised by farmers; and
- establish farmer committees.

IIIT-H

The International Institute of Information Technology, Hyderabad is a research university whose primary goal is to impart a uniquely broad and interdisciplinary information technology education. The Power Systems Research Centre was set up at IIIT-H to undertake research regarding IT applications for power and energy systems. The main activities of IIIT-H in the project were to

- provide assistance for the overall technical approach of the pilot project;
- provide technical knowledge in designing solutions, including the selection of feeders;
- carry out regular field visits, data collection and reporting activities; and
- assess the technical feasibility of measurement and data collection methods used.

SCTI

The Steinbeis Centre for Technology Transfer India provides technology-transfer solutions to concrete local problems in several fields, and its services include technical consultancy, research and development, outsourcing of engineering services and engineering components, training, and international technology transfer. The main activities of SCTI in the project were to

- implement, install, and monitor technical aspects of the project and provide necessary services and materials;
- provide technical background knowledge and training to the field staff; and
- coordinate and conduct regular meetings with farmers to raise technical awareness regarding the project.

cBalance Solutions Hub

cBalance is a knowledge-centric solutions hub that specialises in tool building and strategy development for integrating carbon enterprise resource planning into institutional processes. cBalance was the technical evaluator of the pilot project, with its main activities being to

- develop an overarching strategy for presenting the impact of the chosen measures;
- design a field measurement plan, including development of an algorithm to identify a representative sample of pumpsets and DTRs that would form the target population for the study;
- determine relevant electrical and flow parameters for measurement and demonstration of impacts;
- carry out field measurements to quantity performance impacts of capacitors on a statistically representative sample of pumpsets; and
- undertake data analysis to quantify and assess the scale of the impact achieved as well as to perform rational projections of the impacts throughout the pilot project, based on the sample results.

CWS

The Centre for World Solidarity conceptualised and demonstrated the feasibility of a social regulations approach to sustainable groundwater management, both in drinking water supply and irrigation, from 2004 to 2012 in selected villages in Andhra Pradesh. Based upon this approach, CWS worked with farmers on collective methods to address electricity-related issues in groundwater-based irrigation.

Drawing from these experiences, CWS provided advisory support to the pilot project, for which its activities were to

- deliver inputs for project design, such as the development of implementation plans;
- review implementation processes and reports;
- conduct capacity building among field partners in operationalising plans; and
- provide input to policy intervention processes and events.

PEG

The Prayas Energy Group has been active in the electricity sector in the areas of generation and supply, energy efficiency, renewable energy, and fuels and resources. Power supply for agriculture—in terms of accurate consumption estimates, equitable distribution of subsidies, improved quality of supply and end-use efficiency—has been one of the areas of its work. The main activities of PEG in the project were to

- effectively design project inputs (work plan, field intervention, evaluation, reports);
- provide input for capacity building among field partners; and
- provide input to policy interventions (workshops, regulatory/policy submissions).

6.2 Pilot Project Region

The pilot project took place in the area adjacent to the towns of Vemulavada and Sircilla in the Karimnagar District of Andhra Pradesh (Fig. 6.1). The area is moderately hilly and surrounded by forests, and the Mula Vagu river runs through Vemulavada, though it is dry during most months of the year. The main crops in the area are paddy and cotton, with most irrigation being done by electrical pumpsets from groundwater. Three tanks (i.e., large basins that accumulate rain water during the monsoon season) have been set up to increase the groundwater level. In 2003, a governmental program, carried out by the local NGO SEWS, improved the tanks by constructing small canals to increase water inflow from forest areas. Further, the tanks have been connected with each other and with the Mula Vagu in order to avoid tank overflow and increase water availability. Some villages of the intervention area were already participating in a watershed management program initiated in 2003, which includes savings groups, employment projects, and village committees.

The project area was chosen due to the following considerations:

- Presence of a co-operative society as the distribution utility rather than a larger state-owned utility: A similar project on capacitors had been implemented in

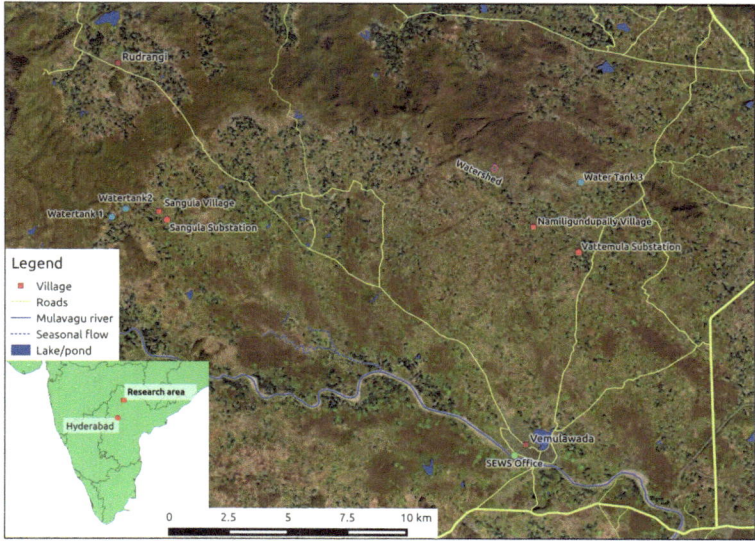

Fig. 6.1 Map of pilot project area. *Source* This map was composed in QGIS 2.6.1 using elements from OpenStreetMap project, Landsat ETM 7 (Bands 1, 2, 3 and 7) and ASTER GDEM (Tile: N18E078). http://download.geofabrik.de/asia/india.html. http://earthexplorer.usgs.gov/. ASTER GDEM and Landsat ETM 7 are products of METI and NASA

an area with such a state utility, and communication difficulties with the utility were reported. Hence, the project team decided to choose an area with a co-operative society that supplied all participating farmers.

- Presence of watershed management activities: Part of the intervention area is subject to a watershed management program funded by the German Development Bank KfW. This brought the advantage of being able to evaluate the effect of capacitors under different conditions.
- Representative cropping pattern: The area is representative for the Deccan Plateau of South India, covering parts of Andhra Pradesh, Karnataka, Tamil Nadu, and Kerala, but especially representative for the Telangana region. The main crops—paddy and cotton—are water intensive, and irrigation is essential for successful harvests (Fig. 6.2). Hence, an interest in and need for improved electrical energy supply was already given.
- Lack of alternative water supply: The area is rather dry, and there is no alternative to groundwater irrigation. A high dependency on good electrical energy quality makes the region a priority for intervention. An added benefit is that the effects of capacitors would be easier to estimate, because confounding with other irrigation methods could already be ruled out.

Fig. 6.2 Paddy field with pumpset near Vemulavada. *Source* Malte Müller

6.3 Aims and Stakeholders

The project aims were developed by taking into consideration all stakeholders simultaneously. The aims were to reduce electrical energy consumption, improve electrical energy quality, reduce line losses, and raise capacity building and awareness among farmers.

The team expected the following benefits for stakeholders: the farmers would profit from improved electrical energy quality, savings in motor and DTR repair costs, and, through improved irrigation and higher yields, gain greater reliability for their planning activities. Further, the DTR committees would be likely to lead to greater "voice" for the farmers.

The utility would be able to reduce the number of units sold to agriculture, which could instead be sold to profit-generating consumers such as industries. Further, better electrical energy quality would likely reduce maintenance and repair costs. Decreased line losses and increased power factor would be an important benefit for annual administrative reports and could serve towards meeting targets set by the Andhra Pradesh Electricity Regulatory Commission.

Indian society in general could expect to benefit from greater availability of electrical energy if agricultural consumption and line losses were to decrease. Furthermore, this may lead to a reduction of CO_2 emissions, if less overall electrical energy is required.

The stakeholders can be differentiated between those directly affected and those indirectly affected by problems related to electrical energy provision for agriculture. Farmers should immediately realise expected effects of project

Table 6.1 Overview of stakeholders involved in pilot project

Stakeholder	Role	Current situation	Change in pilot project
Farmers	Use groundwater for irrigation of farmland	Suffer from high repair costs, low water levels, poor and restricted electrical energy supply, and unfavourable supply timing	Reduction of burnouts; more efficient pumping
Utility	Distributes electric energy; maintains distribution grid	Suffers from high costs of agricultural supply and DTR repair costs	Reduced delivery to agriculture; reduced DTR repair costs
Governments: Andhra Pradesh, Indian	Subsidise electrical energy for utilities; set tariffs and standards	Suffer from high subsidies and economic disadvantages due to limited power supply	Reduced subsidies and supply gap
Electric energy consumers in Andhra Pradesh	Consume electrical energy, pay taxes	Suffer from high taxes for subsidies, power cuts and polluted air	Less power cuts better electrical energy quality

implementation, as results would be already observable in the short run. Also the utility, which is supplying farmers, would be able to sense effects immediately, as reduced electrical energy consumption to agriculture and lower repair costs are realised. The other stakeholders—competing electrical energy consumers, the Andhra Pradesh and Indian governments and their subordinate departments, and, finally, Indian society as a whole—would be affected indirectly and only in the long run. In the following, we will focus only on the main stakeholders and the short-term effects expected from the project.

The interrelations of the stakeholders are summarised in Table 6.1. The team considered the pilot project area to be special in that power distribution is carried out by a co-operative society, CESS (see Sect. 6.1), rather than by one of the four distribution companies operating in Andhra Pradesh. Although CESS fulfils the same duties as the distribution companies do in other areas, there are differences in the subsidisation policy they are subject to. CESS buys each kWh from the distribution companies at a current price of 0.48 INR and sells it at given tariffs to different customer groups. Farmers, however, are supplied at no charge per kWh, so that CESS sustains losses for each kWh delivered to agriculture. In contrast, the distribution companies receive a direct subsidy from the Andhra Pradesh government for units delivered to agriculture, yet buy kWh at a higher price. The incentive structure is therefore different for the distribution companies. The fewer kWh CESS sells to agriculture, the less its losses are. A great incentive for improved energy efficiency is, thus, inherent to CESS. Additionally, CESS is responsible for maintenance of DTRs. Low electrical energy quality on agricultural feeders lead to increased DTR burnouts, a reduction of which would also be in line with the incentive structure of CESS.

Farmers in the region are mostly cotton and paddy farmers. As in other regions in Andhra Pradesh, they suffer from limited water availability and, due to their high dependence on groundwater irrigation via electric pumpsets, from unreliable electrical energy supply. Although farmers do not pay for electrical energy consumption, they would certainly appreciate an improvement in quality, as a high financial burden for them comes from pumpset repair costs. However, due to the subsidy that grants them free electricity, energy efficiency and reduced electrical energy consumption play a minor or indirect role for farmers.

In the long run, the intervention was expected to lead to an increased availability of electrical energy, which could then have been turned into an improvement for the farmers in the form of increased hours of supply. However, many institutional, political and technical hurdles have made this aim unrealistic from the farmers' perspective, and within the geographical scope of the project the effects would have been rather marginal.

6.4 Technical Approach

A low-cost way to increase electrical energy quality is to connect shunt capacitors parallel to the load. The connection of capacitor banks at a substation in combination with capacitors at individual loads seems—based on discussions with electrical engineers—the best way to maintain a power factor of around 0.8–0.9. However, the pilot project aimed towards enabling farmers to implement solutions themselves and, hence, solutions on a larger scale were not pursued. Shunt capacitors are a simple and widely available device, affordable for almost all farmers in the study area, and have exhibited positive effects for both farmers and utilities. Additionally, capacitors can serve as an entry point to investment in other types of energy-efficient equipment.

Capacitors have one further characteristic (see Box 2 in Chap. 2). The effect of a single capacitor is minimal, and only after a certain number of farmers per DTR use capacitors does a significant change become noticeable. This fact implies that (non-)use of capacitors is not only based on technical issues but also on institutional and social settings. Some coordination of the farmers is, thus, required, but missing empirically (Kimmich 2013). Consequently, a project with capacitors not only faces technical challenges but must also take into consideration linkages to the social side of agriculture, thereby increasing difficulty of implementation.

6.5 Social Approach

The social approach taken by the project team consisted mainly of activities performed within farmer committees. Organisation of the farmers took place after the capacitors were installed, in accordance with the following concept. As most

Fig. 6.3 Organisational chart
of farmer committees

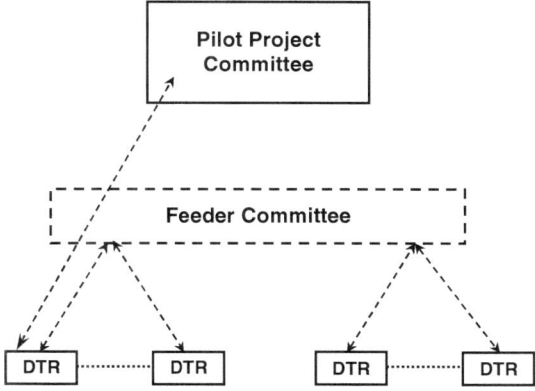

interaction between farmers occurred within one DTR, the smallest unit of a committee was the DTR committee (DTRC). All farmers who were connected to a
DTR thus became members of a DTRC, which were then connected to each other
through a feeder committee (Fig. 6.3).

Elected representatives of the DTRCs—each connected to their respective
feeder—became members of a feeder committee, of which there were two in the
project area. Their main purpose was to solve issues that could not be solved on
the DTRC level. Further they could initiate larger interventions beyond the usual
scope of the project. At a higher level, the feeder committees were then brought
together into the pilot project committee, which served as the head organisation
which communicated with CESS concerning electrical energy supply, solved
internal matters and took important decisions.

Each committee had its own constitution and representatives: a president,
a secretary, a treasurer and, for DTRCs, a technical assistant. There were regular meetings with structured agendas, during which technical and social problems
were discussed and some variables of the motors, capacitors and DTRs documented. The DTRCs met monthly and the feeder committees and pilot project
committee every four months (Fig. 6.4). The representatives kept in contact with
CESS officials and helped their members to solve problems. The technical assistants received advanced training on different aspects of irrigation and electrical
energy and were able to help other farmers with technical problems related to their
pumpsets (A detailed description of the committees and their tasks can be found in
Appendix III).

In general, the social intervention approach facilitated solving the farmers'
problems in a bottom-up and collective manner; helped to avoid conflicts and
misunderstandings between farmers, CESS and the project team; served as a plat-
form for capacity building and training; and assisted interaction between CESS
and the farmers. In the long run, the experience of the committees can be used
as an institutional showcase for collective action. They can also serve as a plat-
form for further interventions. The committee model can be transferred to other

Fig. 6.4 Farmers in front of their DTR. *Source* Malte Müller

fields like water management groups, village electricity committees, and also be used when new technologies, such as solar photovoltaic water pumpsets, are to be introduced.

Reference

Kimmich C (2013) Linking action situations: coordination, conflicts, and evolution in electricity provision for irrigation in Andhra Pradesh, India. Ecol Econ 90:150–158. doi:10.1016/j.ecole con.2013.03.017

Chapter 7
Project Steps in Detail

Abstract This chapter illustrates relevant steps within the pilot project's chronology, the timeframe of which ranged from October 2011 to June 2013. The sections are organised into three chronological phases: preparation and planning, implementation and evaluation.

Keywords Project implementation · Project preparation and planning · Project evaluation techniques · Power factor · Transformer committees

7.1 Preparation and Planning Phase

The preparation and planning phase, including steps required to initiate and facilitate the implementation phase, was finalised after:

- the area for intervention was decided on,
- all partners were made fully aware of the aims and scope of the project,
- an intervention method was agreed upon,
- its technical feasibility was verified,
- local actors were included in the project,
- the project consortium agreed to the implementation strategy, and
- baseline data had been gathered.

The official initiation of the preparation and planning phase took place in October 2011, when two workshops were conducted to decide on the intervention type and location. The following sections describe the main results of this phase.

7.1.1 Rationale for Choice of Intervention

As discussed in the above chapters, the project team decided to install shunt capacitors at the pumpset level to improve the power factor (see Sect. 4.3 for

© The Author(s) 2016
J. Sagebiel et al., *Enhancing Energy Efficiency in Irrigation*,
SpringerBriefs in Environmental Science, DOI 10.1007/978-3-319-22515-9_7

technical details) of agricultural motors in a study area as the main intervention strategy. This decision, the result of intense discussions within the project team and with external experts, was shaped by several constraints. First, the project was to be carried out on a limited budget, which ruled out large-scale interventions such as replacement of motors or DTRs. Hence, the range of choice was reduced to small, low-cost technical solutions. Second, a focus was to be put on farmer participation. Two related concerns were that (a) a feasible project solution needed to be found where the contribution of utilities and regulators was not necessarily required and (b) it should be, so far as possible, unaffected by unexpected political changes and regulatory delays. Third, the solution was envisioned to be technically simple and easy to understand for stakeholders, especially farmers. Fourth, farmers and NGOs should be able to upscale the intervention without the involvement and support of larger actors, like governments. Fifth, the solution should not require intensive maintenance and technical know-how. Sixth, in order to empower farmers, the solution required some scope for collective action and participation.

Considering these criteria, the range of possible solutions was still large: dry-run preventers, for example, would have fulfilled all of these requirements (see Sect. 3.2). Capacitors were principally chosen for the following reasons: first, preliminary but promising achievements from a small project that deployed capacitors as an intervention strategy already existed. These results are summarised in Mohan and Sreekumar (2010); its authors participated with an advisory role in our pilot project (see Sect. 6.1), bringing in their experience on the topic. It turned out that, in their project, capacitors remained installed for several years and were highly valued by the farmers involved, as confirmed via a site visit by the project team in December 2011. It was expected that this successful "pretesting" would increase the likelihood of technically sound and robust implementation in our project. Other possible solutions would have required additional pretesting, which would have been outside the project's scope and finances. Second, the solution of capacitors was recommended within the project team, with the electrical engineers involved confirming their anticipated positive effects on the electric grid. Third, most farmers were already familiar with capacitors and understood how they worked. Fourth, CESS, the local utility, was comfortable with this solution and assured us of its support. Fifth, capacitors were highly recommended by the Andhra Pradesh government and are compulsory in some regions there and India more generally, being standard in many other electrical appliances as well. Sixth, a capacitor is a simple electrical device which has been used and proven beneficial in many appliances for several decades, outruling the risks that are often inherent to new technologies. Meanwhile, seventh, investment and maintenance costs for capacitors are relatively low.

Some potential disadvantages of using capacitors were, however, also clear to us: first, many farmers were reluctant towards capacitor usage, since a government program in 2005 had led to improper installation of capacitors, resulting in adverse effects on pumpset efficiency and reliability (see details in, e.g., Kimmich (2013a) and Sect. 2.3). This hindrance substantiated the need for capacity building and social intervention in order to win acceptance for the project. For the building of trust between farmers and the project team, it appeared crucial to conduct several training sessions to explain how a capacitor works and how to maintain it. Second, as we have already mentioned, capacitors only function properly when concerted action evolves. If only one farmer connected to a DTR uses a capacitor, the effect is rather small. The full potential of capacitors can only be achieved when all farmers connected to a DTR/feeder have installed them (see Box 2 and Box 4). This insight was crucial for planning a comprehensive implementation strategy. It was the task—and major challenge—of the project team, especially the local NGO, to at least achieve full coverage for one feeder, requiring sophisticated grassroots work and capacity building. The project team came to the conclusion that the advantages of capacitors, compared to other technologies, outweigh the disadvantages. Especially the need for collective action was regarded as an under-researched topic which would fit well into the research agenda of the Sustainable Hyderabad project, and the experience and insights gained would likely be important for policy makers working in similar fields.

7.1.2 Selection of Feeders

Having agreed to install capacitors, the project team then searched for an exact project location. Considering the available resources, the project team decided to select two feeders to be completely equipped with capacitors, translating into approximately 800 pumpsets. As the project team had earlier decided to conduct the project in the area served by CESS, only feeders which were under its maintenance were considered. Data on all possible feeders was provided by CESS, and COOP and IIIT-H developed the following criteria to guide the selection of feeders, according to which each feeder should have:

- approval from the local partners: SEWS and CESS,
- a large share of agricultural connections,
- few or no illegal connections,
- a low number of total agricultural connections,
- one watershed and one non-watershed feeder,
- similarities with other selected feeders,
- a representative cropping pattern,
- lines in acceptable working condition,
- no plans for renewal or replacement of the feeder or substation, and
- possibilities for measuring all technical parameters.

Also, one feeder should be located in SEWS's service area. A detailed technical analysis of potential feeders had been carried out by IIIT-H and SCTI after a pre-selection process by SEWS and CESS. CESS based its pre-selection on its own internal preferences, taking into consideration which feeder needed such an inter-

Table 7.1 Situation at the two intervention feeders in January 2012

Feeder	Namiligundupally	Sangula
Domestic load in kW	188	512
Number of DTRs	14	21
Number of connections	346	355
Total kW for agriculture	900.4	991.8
Watershed program	Yes	No

vention most and which one was technically adequate for it. Meanwhile, SEWS preferred that the feeder be in an area where SEWS had already been working, in order to have better access to the farmers and an easier and quicker implementation phase.

Technical analysis of feeder choice was accompanied by field visits and collection of technical parameters. Eventually, a ranking of attributes led to the selection of two intervention feeders: "Namiligundupally" at Vattemula substation and "Sangula" at Sangula substation. The selection process was finalised in January 2012. Table 7.1 lists the characteristics of the selected feeders.

During subsequent project phases, some of the listed characteristics changed and, unexpectedly, the load at the Namiligundupally feeder was heavily reduced, because another feeder from the same substation replaced it for a portion of the overall distribution. The choice of the feeders was an important step, as it enabled the project team to begin making contact with the farmers who would eventually be part of the project. The following steps, thus, deal with our interaction with these farmers.

7.1.3 Social Survey

In order to increase the effectiveness of the implementation phase, the project team decided to collect background information on the participating farmers, including their attitudes and knowledge, socio-demographic variables, cropping patterns, field sizes, currently used technologies, and perceptions regarding groundwater levels. Thus, a social survey was carried out before the farmers were made aware of the upcoming project (Fig. 7.1). The data collection was done in February 2012 and took four days. The sample consisted of 234 farmers from the intervention feeders and from a control group in the surrounding areas. In total, eight villages were covered. A training session of one and a half days with 16 field investigators prepared them to carry out the survey. A discrete choice experiment to elicit preferences regarding different alternative capacitor types and implementation schemes was included in the questionnaire (see Box 3).

The results of the social questionnaire are presented in Table 7.2. In order to assess whether the sample was representative for Andhra Pradesh, the table contrasts our dataset with a similar survey from 2010, which was carried out in

Fig. 7.1 Conduction of social survey. *Source* Christian Kimmich

Table 7.2 Comparison of survey statistics from Andhra Pradesh wide and project samples

	Vemulavada Sample 2012	Andhra Pradesh Sample 2010	Vemulavada Sample 2012		Andhra Pradesh Sample 2010	
A. **Farm/ Household Variables**	**Mean (standard deviation)**	**Mean (standard deviation)**	**Min.**	**Max.**	**Min.**	**Max.**
Acres irrigated during *kharif* season (April–October)	3.68(2.85)	3.66 (4.45)	0.08	21.5	0	56.5
Additional household income (total and share in %)	8249.85 (15,370.48)	65	0	150,000		
Participation in agricultural training (share in %)	32.70	33				
Participation in the *Gram Sabha* (village meeting) (share in %)	60.36	54				

(continued)

Table 7.2 (continued)

	Vemulavada Sample 2012	Andhra Pradesh Sample 2010	Vemulavada Sample 2012		Andhra Pradesh Sample 2010	
Member of a farmer association (share in %)	52.89	22				
Education (years)	7.81 (3.44)	3.80 (5.06)	1	17	0	18
Age (years)	47.34 (13.05)	44.34 (13.58)	20	78	19	83
Gender (share of male farmers in %)	85.00	81				
Caste (share of scheduled caste/scheduled tribe farmers in %)	30.96	35				
Household size	5.14 (2.27)	5.55 (2.71)	2	16	2	19
B. DTR variables	**Mean (standard deviation)**	**Mean (standard deviation)**	**Min.**	**Max.**	**Min.**	**Max.**
DTR burnouts per year	5.66 (4.04)	1.02 (1.04)	1	16	0	7
Farmer's cost for DTR repair (INR)	642.41 (1540.12)	620.58 (869.65)	30	10,000	0	8000
Number of farmers connected to a DTR	9.30 (4.92)	17.30 (8.12)	1	18	1	50
Costs for authorization of connection (INR)	9,209.61 (8430.64)	7180.11 (8742.22)	250	1,000,000	0	100,000
Bribes paid for receiving connection (INR)	1,988.75 (2909.87)	946.60 (1456.48)	50	300,000	0	10,000
Farmers with DTR head position (share in %)	27.78	69.00				
Farmers owning their DTR (share in %)		3.00				

(continued)

Table 7.2   (continued)

C. **Pumpset variables**	Vemulavada Sample 2012 Mean (standard deviation)	Andhra Pradesh Sample 2010 Mean (standard deviation)	Vemulavada Sample 2012 Min.	Max.	Andhra Pradesh Sample 2010 Min.	Max.
Motor burn-outs per year	2.03 (2.19)	1.86 (1.64)	0	20	0	12
Costs for motor repair (INR)	4147.5 (3216.54)	2693.15 (1513.11)	800	30,000	200	8500
Age of pumpset (years)	16.19 (9.90)	7.21 (5.94)	0	42	0	30
Branded pumpset (share in %)		53.00				
ISI-marked pumpset (share in %)	63.30	37.00				
BEE-rated pumpset (share in %)		6.00				
Capacitor successfully installed (share in %)	9.76	10.00				
Automatic starter installed (share in %)	40.27	85.00				
Pumpset investment costs (INR)	14,499.51 (7945.12)	22,342.90 (8998.48)	1,500	56,000	2000	72,000
Pumpset maintenance costs (INR/year)	690.16 (997.26)	468.11 (375.09)	0	12,000	50	3000
Well depth (feet)	24.67 (26.21)	166.79 (69.82)	4.57	200	13	400
Well investment costs (INR)		23,324.51 (18,647.77)			1000	150,000
Months without water		4.91 (1.60)			0	7
Well runs dry in summer (share in %)	83.98	95.00				

Source Adapted from Kimmich (2013b) and own material

multiple districts in Andhra Pradesh with the aim to better understand dilemma situations in Indian agriculture (Kimmich 2013b). Part A of the table represents socio-economic variables from both datasets. The data from the project sample is similar to the data from Kimmich (2013b), with the main differences only being in terms of membership in a farmers' association and average number of years of education. The average farm size was 3.68 acres per farmer, with a range from 0.08 to 21.5 acres, indicating some heterogeneity among them. Part B of Table 7.2 reports variables at the DTR level. The number of DTR burnouts per year was relatively high in the study area, with nearly six burnouts, compared to the 2010 data, with only about one burnout. Furthermore, the average number of farmers connected to one DTR was low compared to the data from 2010, reflecting the higher share of head-position farmers at the DTR. The number of farmers owning their own DTR was only surveyed in 2010 and is expectedly low (3 %). Part C displays pumpset variables. Most relevant is the yearly average of around two motor burnouts per pumpset. Combining this figure with an average repair cost of 4,147.50 INR (or 2,693.15 INR in 2010) demonstrates the high cost of inappropriate technologies and low electrical energy quality borne by farmers. The number of motor burnouts is not significantly different between the samples. Only 10 % of the farmers from both surveys had already successfully installed a capacitor into their pumpset, confirming the already existing reluctance towards capacitors described in Sect. 7.1.1.

Box 3: Using Discrete Choice Experiments to Calibrate the Pilot Project Intervention

This box is based on unpublished research of the authors.

Discrete choice experiments have been extensively used to evaluate goods, services or policies in transportation, environmental, and health economics as well as in marketing. In development economics, the method has been applied for eliciting preferences in sectors like food, water or farming to provide coherent policy recommendations (Bennett and Birol 2010). The widespread use of discrete choice experiments is due to their rather simple yet general application format combined with robust underlying economic theory.

More specifically, a discrete choice experiment is a survey-based, stated-preferences method in which respondents are asked to repeatedly choose between alternatives. Each alternative is described by attributes, which vary from choice task to choice task. One advantage of this method, compared to other stated-preferences methods, is that it can enable evaluation of the attributes of a good or policy rather than the good or policy itself. For example, a policy maker is intent on implementing a sustainable wetland management policy. He knows that there are different—maybe equally expensive—variants of the policy in terms of parameters such as degree of biodiversity or size of open-water surface area, yet is unaware of the preferences of the local population. Further, he might be interested in quantifying the perceived value of the different variants to compare them to related costs. Using discrete choice experiments can be a means for identifying a "socially optimal" policy by adjusting implementation to preferences. The example described here (Birol et al. 2006) is one of many such experiments that have been conducted to guide policy makers in environmental policy questions.

For the pilot project in Andhra Pradesh, we applied a discrete choice experiment for a very different purpose. Here, the project team wanted to identify the preferences of potential beneficiaries (i.e. a small subset of people who would be directly affected) in order to adjust the implementation procedures of small pilot projects or development cooperation projects. The main difference between classical applications and ours lies in the target group and generalization of results. Such experiments usually seek to be relevant for large groups or whole populations (e.g., marketing a new product, constructing a new highway, preserving a natural habitat, setting up a new health insurance scheme) and are aimed at attaining general conclusions. Meanwhile, our application was restricted to a specific and unique intervention, generally guided by the following rationale: Whenever new ideas are to be tested in the field, some fine tuning is necessary. The fine tuning varied from case to case. When preliminary information-gathering is not conducted beforehand, trial and error costs tend to be higher than necessary. While many methods for this exist and are frequently applied, discrete choice experiments appear to offer some outstanding benefits. First of all, a high degree of realism is provided, as different scenarios are presented to the respondents. Second, the comparative nature of discrete choice experiment tasks makes decisions easier for respondents, compared to surveys where direct statements of, for example, willingness to pay are required. Third, the method is relatively efficient in that very precise and quantifiable findings can be collected in a relatively compressed format. Fourth, with good sample selection and proper statistical design, estimates tend to be reliable, as it is easy to cover large shares of a target population. Fifth, a discrete choice experiment task is likely to be more compelling to respondents than answering simple, perhaps boring, questions.

The story of the pilot project's discrete choice experiment is as follows: after interviewing several farmers and experts, the project team still lacked precise and representative information on the farmers' preferences regarding different capacitor attributes, especially non-technical factors such as capacitor warranty and costs. Further, we needed to determine whether farmers would be willing to join the distribution transformer committees (DTRC). In the end, it all depended on the willingness of the farmers to participate or not, because if they were not willing to adopt the proposed changes, the project would likely fail. Sometimes, even small deviances from a preferred application can prevent successful implementation. In order to inhibit this project from falling into such a trap, it was deemed essential to find out

- whether farmers were at all willing to install capacitors;
- whether they were willing to pay a price to acquire a capacitor;
- whether all farmers would be interested in joining the DTRC;
- to what extent a warranty would be appreciated by farmers; and
- how warranty, price and the DTRC could be traded off and interact.

It was felt that finding answers to these questions could help to guide implementation of the project, especially in possibly predicting what would "work" or "fail". For example, if we had found out that farmers were willing to pay for capacitors only when they came with a warranty, it would have been fatal to provide capacitors at a charge but without warranty. Further, it was useful to investigate how farmers traded off between the price and warranty, meaning how much a farmer would pay additionally for one year of warranty.

The idea was put into practice in a survey with 234 farmers from the intervention villages and from neighbouring "control group" villages in February 2012. The discrete choice experiment consisted of three alternatives: "no capacitor", "only capacitor" and "capacitor and cooperative" (cooperative being the word used for DTRC in the survey). The latter two alternatives comprised two attributes, each with two levels which appeared in different combinations between choice situations. The attributes were "years of warranty" (level 1: no warranty /level 2: five years warranty) and "investment costs for capacitor" (level 1: no costs /level 2: 300 INR). Figure 7.2 presents one out of the 16 choice situations which were used in the survey. Each participant responded to eight choice situations after a detailed explanation of the alternatives and attributes. The results shed light on the concerns raised above.

The analysis was carried out by applying different methods. First, simple decision heuristics were detected, meaning where a respondent always chose the same alternative, regardless of the levels of the attributes or he always chose the alternative where one specific attribute was better. From Table 7.3 it can be seen that only one respondent always opted against capacitors, 15 respondents always chose the alternative "capacitor and cooperative", 26 respondents always chose the alternative that was cheaper and 39 respondents always opted for longer warranty. Another 43 respondents chose either based on the cost or the warranty attribute. This analysis indicated that farmers are not in general reluctant towards capacitors, even when they cost something and come without warranty. Further, more respondents were keen on having a longer warranty than on free capacitors, revealing a willingness to pay for capacitors.

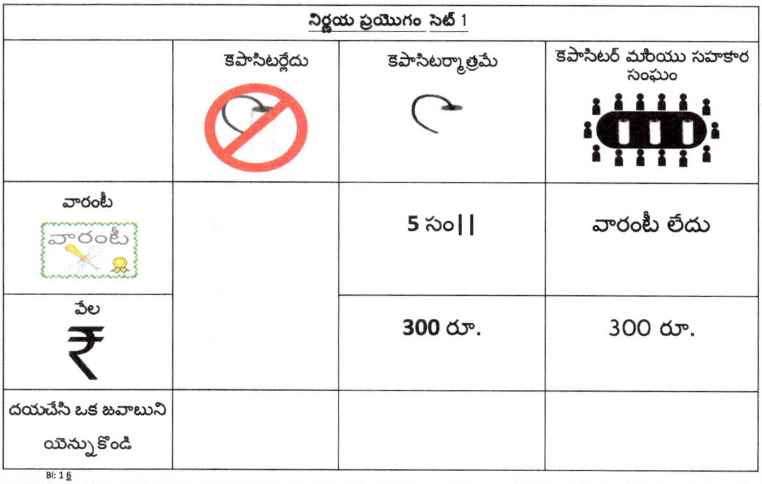

నిర్ణయ ప్రయోగం సెట్ 1		
	కెపాసిటర్లేదు	కెపాసిటర్మాత్రమే	కెపాసిటర్ మరియు సహకార సంఘం
వారంటి		5 సం||	వారంటీ లేదు
ఫేల ₹		300 రూ.	300 రూ.
దయచేసి ఒక జవాటుని యెన్నుకొండి			

BI: 1 &

Fig. 7.2 A sample choice situation from the survey; each farmer was asked to respond to eight of such situations

Table 7.3 Simple choice heuristics farmers have used in the discrete choice experiment

Always choose…	Frequency	Percent (%)
"No capacitor" alternative	1	0.43
"Only capacitor" alternative	2	0.85
"Capacitor and cooperative" alternative	15	6.41
The cheaper alternative	26	11.11
The alternative with longer warranty	39	16.67
Either cheaper price or longer warranty	43	18.38
Not in any pattern	108	46.15

The data was further analysed with microeconometric methods. Applying a conditional logit model, we calculated willingness to pay values (i.e., trade-offs between the attributes and the cost) for the attributes. The model estimates in terms of willingness to pay are presented in Table 7.4. The sampled farmers were more likely to choose the alternative with a cooperative and—not surprisingly—preferred lower costs and longer warranties. They were on average willing to pay an additional 325 INR to join the cooperative and an additional 117 INR for one more year of warranty. Although the analysis could have been widely extended, such as by investigating preference heterogeneity or correlating the choices with socio-demographic variables, important insights were already obvious and were directly useful for project implementation.

Table 7.4 Conditional logit regression results expressed in willingness to pay conditional for those choosing an alternative with capacitor

Attributes	Willingness to pay in INR	95 % confidence interval	
		Lower bound	Upper bound
Cooperative	324.9[***]	243.9	405.9
One year Warranty	117.4[***]	89.7	145.1
Costs	−1	−1	−1
Number of observations	3550		
Number of respondents	233		
Count R^2	0.735		

*** significant at the 1 % level

The advantages of such precise and extended analyses come at the cost of extra effort. A discrete choice experiment, compared for example to only using focus group discussions, should only be used if the additionally generated information really seems essential for successful implementation. Arguments against the use of discrete choice experiments include the following: first, discrete choice experiments involve much preparation and are relatively cost intensive. Like most quantitative methods, a relatively large sample size is required to obtain statistically sound estimates, and field investigators need to be hired and trained for at least one full day. Second, falsely selected or omitted attributes or a poor experimental design can bias the results and lead to inaccurate or misleading implications. Third, due to the hypothetical nature of the method, respondents might not answer in the same manner that they would in real decision situations (hypothetical bias).

The observed low number of installed capacitors may be partially attributed to a coordination failure that has not provided any incentives for farmers to install capacitors on the grid—a core coordination problem that has already been explained theoretically in Box 2. Box 4 outlines an experiment carried out in the pilot project region in February 2013 project with farmers that have not participated in it to study the dynamic complexity that connected farmers face when installing capacitors.

Altogether, the survey confirmed for the project team the anticipated problems with electrical energy supply among farmers in the region, even indicating that, compared to Andhra Pradesh as a whole, they were facing even greater problems.

Box 4: Applying a Framed Field Experiment to Study Farmers' Reluctance to Install Capacitors

This box is adapted from Müller and Rommel (2013)

Interventions to foster the installation of capacitors have often failed in the past. An underlying coordination problem (as reckoned in Box 2) may explain this reluctance. Contextual factors, such as the number of farmers connected to one DTR, a lack of insight among leading farmers guiding the decisions of fellow farmers, or heterogeneity among farmers might aggravate non-adoption. We studied the role of group size and leadership on the capability of farmers to overcome this coordination failure by means of a framed field experiment.

The experiment examined the s-shaped production function of electricity quality at DTRs in the study area (Fig. 7.3). Its shape creates a coordination problem among investing farmers: it is individually rational to install a capacitor only when the slope of the production function is relatively steep (area 2), whereas a farmer should stay away from buying a capacitor when the slope is flat (areas 1 and 3; cf. Kimmich (2013a)). Individual investment decisions depend, therefore, on the behaviour of others. In reality, farmers are predominantly trapped in area 1, where no one would invest—a coordination failure.

In each out of 12 rounds of the experiment, participants decided to buy or not to buy a capacitor. Participants played either in groups of five or groups

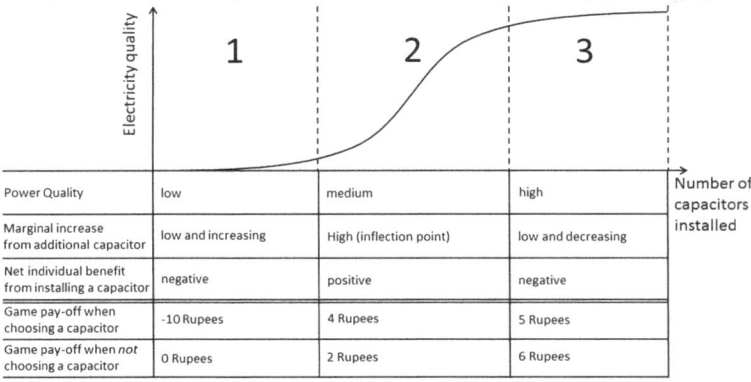

Power Quality	low	medium	high	Number of capacitors installed
Marginal increase from additional capacitor	low and increasing	High (inflection point)	low and decreasing	
Net individual benefit from installing a capacitor	negative	positive	negative	
Game pay-off when choosing a capacitor	-10 Rupees	4 Rupees	5 Rupees	
Game pay-off when *not* choosing a capacitor	0 Rupees	2 Rupees	6 Rupees	

Fig. 7.3 Schematic representation of the capacitor adoption coordination problem. *Source* Adapted from Kimmich (2013b)

of ten. In six of the rounds, they made their decisions in parallel (simultaneous treatment), while in the other six rounds a randomly selected participant—the "leader"—moved first (leadership treatment).

In addition, payoffs were also altered to test subjects' sensitivity towards changing financial incentives. The game was played with 225 farmers in eight sessions in different villages close to the pilot project area. Results indicate that participants in smaller groups chose slightly higher shares of capacitors and earned more compared to participants in larger groups. Extrapolating this effect to group sizes of 30 farmers, as they occur in reality, could show that smaller groups are much more able to coordinate their investments into capacitors efficiently than larger ones. Introducing leadership, however, triggered a negative effect. It turned out that having a lead farmer caused farmers to choose less capacitors compared to simultaneous play, which might be partially explained by the possibility of free-riding in area 3 of the curve. Increasing financial incentives, by increasing payoffs, led to higher shares of capacitor adoption (around 12 % of total choices), indicating that a subsidy for capacitors would likely increase real adoption rates.

7.1.4 Technical Survey

Parallel to the social survey, a technical survey was conducted prior to the implementation phase (Fig. 7.4). To elicit the potential of capacitors, the survey included readings before and after connecting capacitors to the motor circuits with about nine sample pumpsets. All parameters, including power factor, before and after connecting capacitors were subsequently compared. The average observed improvement in power factor reached 14–15 %.

The testing helped in the selection of the right capacitor type and capacitance and gave a first impression of the effects that could be expected. In order to get an idea of the exact number of required capacitors, a census of all pumpsets operating under the Namiligundupally and Sangula feeders was carried out. Four local electricians—trained and equipped with meters, tools and safety gear by SCTI staff—were employed to measure relevant parameters and document the types and condition of the motors and pumpsets. When taking measurements at the DTRs, linemen from CESS were included in the process to assure safety standards. The technical survey continued during the installation of the capacitors and, after finalisation of the installation phase, SCTI took 60 more measurements. Following completion of the technical survey, all required data for the implementation phase, described in the next section, was collected.

Fig. 7.4 Measurement during technical survey. *Source* Christian Kimmich

7.2 Implementation Phase

After finalising the intervention strategy and generating the baseline social and technical data, the project team initiated the implementation phase, which began with the selection and purchase of capacitors along with the conducting of awareness-raising meetings for farmers. The meetings were intended to facilitate the persuading of farmers, who had not been made aware of the project until then, about the benefits of the project and convince them to cooperate with the technical team. The implementation phase also included the formation of farmer committees, training sessions and meetings with CESS and the farmers. As mentioned earlier, during implementation some capacitors heated up, caught fire and destroyed the starter boxes of the motors to which they were attached. This led farmers to uninstall the capacitors, after which the project team decided to replace all capacitors with a better model. The following sections explain all of these points in detail.

Fig. 7.5 Capacitor used
in the pilot project. *Source*
Malte Müller

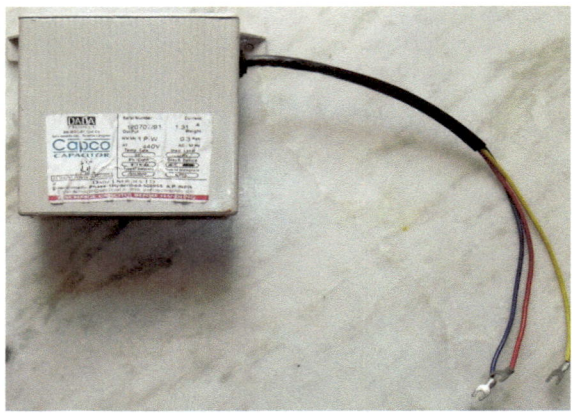

7.2.1 Initial Capacitor Selection

In making the initial selection of capacitor type and make, we considered capacitors from manufacturers such as Havells, Ikon, Neptune Ducati, Shreem and CAPCO. The capacitor make was chosen based on the following factors:

- possibility of earthing at site,
- manufacturer reputation,
- company's service and expertise in agriculture,
- price,
- warranty,
- size and ease of fitting,
- whether manufacturer could provide technical assistance or not, and
- on-site test results (see Sect. 7.1.4).

In the end, the basic model from CAPCO was chosen, as it seemed to be the most appropriate product with respect to the above criteria (Fig. 7.5).

Table 7.5 shows the number of capacitors required at the selected feeders and the resulting costs, based on the technical survey.

7.2.2 Farmer Awareness-Raising Meetings

Prior to the capacitor installation phase, the project team tried to raise awareness about the project among farmers, with SCTI and SEWS conducting several meetings with them. Topics included explanation of the pilot project, involved actors, the goals behind capacitor installation, maintenance issues, and potential advantages to the farmers. To make the project more comprehensible to the farmers, SCTI and SEWS printed leaflets in the local language, Telugu, and distributed

Table 7.5 Summary of technical survey feeder wise

Motor capacity (hp)	Total number of motors	kVAR to be connected	Cost per kVAR in INR	10 % additional, spare capacitors	Overall cost in INR
Sanugula feeder					
3	334	1 kVAR	85.05	33	31,213
5	211	2 kVAR	78.75	21	36,540
7.5	1	3 kVAR	78.75	1	709
10	4	4 kVAR	78.75	1	1890
15	3	5 kVAR	78.75	1	1969
	553			**58**	**72,321**
Namiligundapalli feeder					
3	288	1 kVAR	85.05	29	26,961
5	58	2 kVAR	78.75	6	10,080
7.5	1	3 kVAR	78.75	1	473
10	1	4 kVAR	78.75	1	394
15	1	5 kVAR	78.75	1	788
	349			**38**	**38,695**

them within the villages. Officials from CESS who had not been involved in the planning phase were also made aware of the program. The farmers generally responded positively, though occasionally mentioning doubts. For instance, some farmers pointed to previous capacitor projects, where their installation had hindered the starting of motors. Most doubts were cleared up by the project team explaining the reasons behind the failures, such as faulty installation, and how such problems were to be avoided through the project's approach. The meetings took place from February 2012 onwards, continuing on a regular basis, even during the installation process.

7.2.3 Capacitor Installation

The same electricians who had conducted the technical survey also carried out the capacitor installation, with SCTI monitoring the whole process. The electricians began installation at Namiligundupally feeder, as farmers there had been more easily approachable because of earlier interventions by SEWS. After completion of Namiligundupally, the Sanugula feeder was covered. Spare capacitors, about 10 % of the total purchased, were stored locally, to be used for replacement of damaged capacitors and in case new pumpsets were connected to the feeder.

Fig. 7.6 Installation of
a capacitor. *Source* Malte
Müller

The electricians mounted a capacitor to a wooden or metal panel inside of each
motor starter box and fitted the three phase wires of the capacitor in parallel to
the starter (Fig. 7.6). Installation took place between August and November 2012.
During the process, problems in the field occurred which led to a delay of one
month, including difficult access to some starter boxes, weather conditions during the monsoon season, uncooperative behaviour of some farmers, incorrect rating of some motors due to rewinding and continuous changes in motor rating due
to replacement.

The electricians collected and recorded relevant technical data regarding the
motors and DTRs with the help of linemen from CESS, both before and after connecting the capacitors. The technical staff of SCTI supervised the electricians and
ensured timely completion of tasks. The resulting information was then shared
with all project partners. Further, SCTI made sure that all necessary technical
equipment was available and that employed staff received salaries on time, while
also solving unforeseen problems, such as with unsatisfied farmers.

Having installed the capacitors, the social team started on the formation of
farmer committees.

7.2.4 Establishing Farmer Committees

As explained above, one distinctive feature of the project was to establish an institutional structure through which farmers could be organised and have the possibility to act collectively as members of a committee (see Sect. 6.5). Therefore, to
establish such committees was a key task during the implementation phase, and
their formation was begun in December 2012, after all capacitors had been
installed. First, DTRCs were formed (Fig. 7.7). The project team developed a
detailed work plan and trained a team of three social mobilisers and a supervisor
recruited from the local area. The formation of each DTRC took three days. The
social mobilisers used the first two days to mobilise and prepare the farmers, forming the DTRC on the third. Each farmer of a DTRC signed an agreement (see
appendix IV), including conditions such as the farmer agreeing to invest two hours
per week of voluntary work and contribute a monthly fee of about 100 INR to
cover expenses for meetings. After signatures had been taken, the committee
supervisor explained the contents of a constitution, formulated by SEWS, to the
farmers.[1] The first meeting was then conducted according to the Minutes of the
Meetings log book (see appendix III), which consisted of the constitution and
seven agendas for that meeting, providing a structure for it and some space to
record major points of discussion as well as a chapter to collect some relevant statistics, including number of burned motors within the DTR area.

By the end of December 2012, all DTRCs had been established. The following two months were then used to form feeder committees and to conduct further
DTRC meetings. During the initial meetings, the technical team provided training sessions to interested farmers, explaining the main features of capacitors and
pumpsets and how to maintain them properly. Most training sessions ended in discussions among the famers, often about the poor overall electrical energy supply
situation that they faced.

In addition to attending regular meetings, the head farmers of the DTRCs paid
a visit to a village where capacitors had already been used for more than ten years,
discussing them with the local farmers and looking at pumpsets with installed
capacitors. This exposure visit appears to have created trust towards the technology and helped the farmers better understand the potential long-term effects of
capacitors (Fig. 7.8).

The social team accompanied the committee meetings until the end of the project. Farmers seem to have been satisfied with the procedure overall, but occasionally the social team faced difficulties in motivating them to participate in the
meetings.

---

[1]For illiterate farmers, a project team member read out the constitution and a fingerprint was used
as a substitute for the signature.

Fig. 7.7 Farmers and social team conducting the first DTRC meeting. *Source* Malte Müller

Fig. 7.8 Farmers interacting during exposure visit. *Source* Julian Sagebiel

7.2.5 Cooperation with CESS

Throughout all phases of the project, CESS had been made aware of the components and processes involved and assisted SCTI and SEWS with local and technical requirements. Further, the project team aimed to bring farmers and CESS together into a working relationship characterised by mutual cooperation and understanding. Hence, apart from regular meetings between the CESS

Fig. 7.9 Metering at a starter box, together with the CESS Managing Director (foreground left with glasses). *Source* Malte Müller

Managing Director and the project team, field visits were also carried out where the Managing Director and his assistants visited the studied villages, took motor measurements (Fig. 7.9) and discussed problems related to electrical energy with the farmers. The first visit, to Namiligundupally, took place in November 2012 and resulted in extensive discussions between CESS and the farmers, who mentioned various problems they faced with their electrical energy supply; the Managing Director assured that CESS would resolve them. After this meeting, the project team encouraged farmers to write up letters to CESS, explaining the problems of the DTRCs. A template was provided to the farmers, who wrote their letters during DTRC meetings. The project team then collected the letters and brought them to CESS headquarters in Sircilla.

The second meeting took place in February 2013, serving as the initiation of implementation phase II (see next section). Again, the meeting took place in Namiligundupally. In light of the problem that had arisen with the original capacitors, new capacitors were introduced by the project team, together with the capacitor manufacturer, CAPCO.

Apart from engaging in the above-described interactions with farmers, CESS and the project team agreed to work towards formulating a contract between CESS and the farmers. To incentivise long-lasting installation and maintenance of capacitors by farmers, CESS agreed to discount their monthly electricity fees by 30 INR and an improvement of services from CESS, if farmers agreed to maintain their capacitors and take better care of the overall grid at the DTR level by performing tasks such as tree cutting and better maintenance of motors. To enable farmers to comply with the contract, training sessions for them were conducted by electricians in agreement with CESS.

7.2.6 Major Issues of Phase I

As already mentioned, technical problems emerged in February 2013, as high voltage fluctuations led to an overheating of the capacitors. In some instances, a capacitor caught fire and destroyed the whole starter box (Fig. 7.10). As this phenomenon happened more frequently, farmers started to disconnect the installed capacitors. The technical reasons for these occurrences were never fully understood, so production faults could not be ruled out. The project team decided to replace all capacitors with an improved version. The manufacturer, CAPCO, agreed to distribute new capacitors at no cost and, in collaboration with the project team, designed a new box in which the capacitors were to be placed.

The new box included a miniature circuit breaker and three indicator lights for each phase, to signal the health of the capacitor (Fig. 7.11). The project team decided to continue with the project only at the Namiligundupally feeder, where farmers were more cooperative and because the grid quality there was better

Fig. 7.10 Burned out starter panel. *Source* Malte Müller

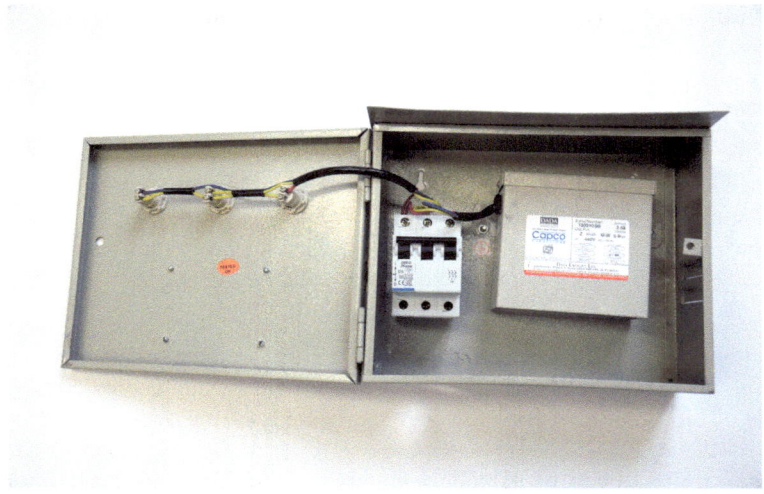

Fig. 7.11 Redesigned capacitor box with miniature circuit breaker, used in phase II. *Source* Malte Müller

than at Sanugula. Subsequently, the project duration was extended to June 2013, instead of February 2013, with all partners continuing their work to completion. Moreover, the evaluation team conducted another round of measurements with the new capacitors to make sure that the expected effects were still present and reportable.

Installation of the new capacitors began in March 2013 and was completed by the end of April 2013. During this time, the social mobilisers continued their efforts with DTRC and feeder-committee meetings and, in order to compensate the farmers further, the project initiator employed a new team of professional trainers to provide training sessions to interested farmers. The whole process received positive feedback from the farmers and from CESS.

7.3 Evaluation Phase

7.3.1 Rationale

To capture sought-after technical effects, cBalance initiated the evaluation phase during the implementation phase: conducting initial measurements by the end of November 2012 and finalising the data collection phase in April 2013. The same measurement approach was used for both capacitor models. A team of two researchers from cBalance and four local electricians took measurements of several technical parameters to estimate the performance of capacitors and water flow rates. Specifically, the evaluators focused on studying the following impacts related to the adoption of capacitors on overall electrical system performance associated with agricultural pumpsets:

- power factor at the pumpset and DTR levels,
- reactive power at the pumpset and DTR level,
- voltage level and imbalance at the pumpset and DTR level,
- total apparent power at pumpset and DTR level,
- current (amperes) required to operate a given active power load, and
- flow rate delivered by pumpsets under identical conditions of power drawn from the grid.

The estimated parameters were then used as a basis for a comprehensive analytical study which sought to ascertain the energy efficiency and energy conservation impacts of the capacitors and the economic benefits of the achieved energy efficiency for the utility.

Technical data sheets of individual pumpsets and DTRs before and after connecting the capacitors were compared (see Appendix II), in order to observe improvements in voltage and power factor. The capacitors had been connected in different ways, as detailed in the next section. Since reduction in kVA of a motor would imply a reduced load on the DTR, the percentage change in kVA was calculated as follows:

$$(kVA1 - kVA2) * 100/kVA1$$

where $kVA1 = $ kVA before connecting the capacitor and $kVA2 = $ kVA after connecting it.

7.3.2 Technical Evaluation Methods

cBalance developed three primary options for studying the relative impact of capacitors on improving energy performance—higher voltage, lower reactive power, higher power factor—at the pumpset level:

1. Control group method: parallel comparison of energy-performance parameters across a control group of pumpsets and motors with no capacitors installed and a similar population of 'intervention' pumpsets and motors with capacitors installed.
2. Sequential method: measuring the energy-performance parameters of pumpsets in the intervention group across a chronological series of events, during which capacitors are successively connected in a pre-determined order across the network. An improvement in energy performance is then measured across the network to establish the incremental impact of capacitors.

3. Batch-wise method: comparison of energy-performance parameters through batch-wise connection and disconnection of capacitors connected to pumpsets networked to a specific DTR, in order to demonstrate a 'before' and 'after' impact at the DTR level while maintaining overall system equilibrium.

Since technical effects were presumed to be immediately observable, options 2 and 3 were preferred. Option 2 would enable measurement of overall system improvement as well as the incremental impact of a single capacitor on improvement of power factor at a single pumpset motor. However, guaranteeing stable and comparable hydraulic pumping conditions over the extended period of time required for such an evaluation is challenging and often precludes this option. In the following, options 2 and 3 are explained in greater detail.

7.3.2.1 Sequential Method

It was hypothesised that capacitors connected to a single DTR will exhibit an 'incremental' benefit, which means that adding a capacitor to a single motor not only affects the energy-performance parameters of that motor but also affects the parameters of other motors and pumpsets connected to the DTR (Kimmich 2013a). Furthermore, with the successive addition of more capacitors, the effect of single capacitors on other motors' energy-performance increases. Validating the hypothesis and measuring the incremental and synergistic interaction of capacitors was the central emphasis when applying this evaluation method. Capacitors were added sequentially to the network in a predetermined order, and the measurement of energy-performance parameters of successive motors was conducted in a chronological manner. The essential condition for this evaluation method was that all motors at the DTR should be running unimpeded during the entire time span of measurement, with measurements being taken at both the pumpset and DTR levels.

Energy-performance parameters of low-tension transmission lines at the DTR level were recorded simultaneously to the readings of each pumpset motor. By application of a fixed energy meter, readings at the DTR level were taken every 30 min throughout the evaluation period to measure the effect of a single capacitor on overall system performance; this timing was arrived at by considering that the performance of a motor was observed to take approximately 30 min to stabilise after connection of a capacitor and restarting of the motor.

Readings at the pumpset level were conducted in the following way:

1. connect the power meter appropriately to the starter box.
2. turn on the motor and wait for 10 min to warrant system stability.
3. take readings of all three phases.
4. turn off the motor and disconnect the capacitor.
5. turn on the motor and wait for 10 min.
6. take readings after an interval of 5 min.
7. repeat the previous step two times.

Readings were taken both with and without capacitors to evaluate the change in energy performance parameters that occurred due to capacitor installation.

Altogether, this evaluation method enabled the determination of the

- overall improvement in the energy performance parameters of the system by measuring change in the parameters between the time when no capacitors were installed in the network and when all capacitors were installed;
- incremental effect of capacitors stemming from the successive addition of capacitors in the system by measuring change in the energy performance parameters of a single motor; and
- effect of a single capacitor and the incremental effect of capacitors on overall system performance by measuring the energy performance parameters simultaneously at the DTR and pumpset levels.

7.3.2.2 Batch-Wise Method

Complementary to the incremental approach, a batch-wise measurement was used to test the aggregated impact of all capacitors on the overall performance of the DTR. In the 'before' condition, multiple measurements were taken at the DTR level, with all capacitors disconnected from their associated motors. Subsequently, in the 'after' condition, the same readings at the DTR level were taken with all capacitors connected to the motors. With this approach, it was crucial to make sure that all motors were running during the entire measurement timeframe.

The chronological order of events using this method was

Step 1: Disconnection of all capacitors in a given DTR network and taking multiple readings after 10 min at the DTR level

Step 2: Connection of all capacitors in a given DTR network and taking multiple readings after 10 min at the DTR level

The major advantage of this method compared to the incremental approach was a relatively low expenditure of time and resources. However, assessing the incremental effect of capacitors on the system and incremental effect of capacitors on the other motors connected to the same DTR was not possible.

7.3.2.3 Comparison of Methods

The basic ideas of both methods of evaluation and their relative advantages are summarised in Fig. 7.12. Although the incremental and the batch-wise method were used in the pilot project, the latter one was the evaluation method selected as being the most suitable and reliable means for assessing the overall impact of the intervention throughout the study. This method led to the most stable 'before' and 'after' conditions and a higher degree of system equilibrium during the measurement time period, compared to the sequential method. Since relatively constant

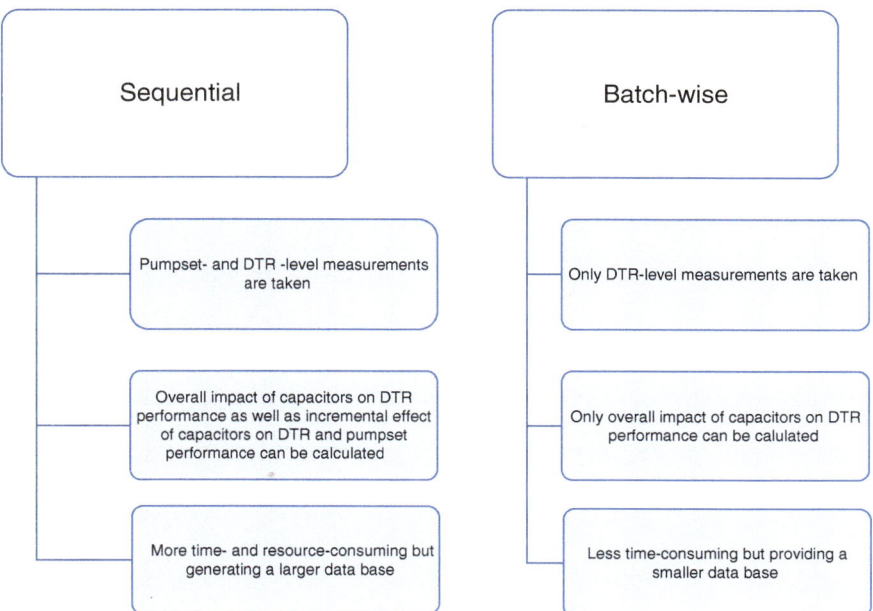

Fig. 7.12 Comparison of sequential and batch-wise method

pumping conditions had to be ensured for only a 48-h period in most cases, this was easier to achieve as opposed to the week-long study periods required per DTR using the other methods. The results obtained through the batch-wise method are, therefore, the primary focus of the evaluation presented in this SpringerBrief. The results of using the incremental method are documented in an internal project report (Gilani 2013) and are available on request from the authors.

References

Bennett J, Birol E (eds) (2010) Choice experiments in developing countries: implementation, challenges and policy implications. Elgar, Cheltenham

Birol E, Karousakis K, Koundouri P (2006) Using a choice experiment to account for preference heterogeneity in wetland attributes: the case of Cheimaditida wetland in Greece. Ecol Econ 60:145–156

Gilani V (2013) Technical evaluation of energy conservation and GHG emissions reduction from capacitor installation on agricultural pumpsets, unpublished project report

Kimmich C (2013a) Networks of coordination and conflict: governing electricity transactions for irrigation in South India. PhD Dissertation, Humboldt-Universität zu Berlin. Shaker, Aachen

Kimmich C (2013b) Incentives for energy-efficient irrigation: empirical evidence of technology adoption in Andhra Pradesh. Energy Sustain Develop, India. doi:10.1016/j.esd.2013.02.004

Mohan R, Sreekumar N (2010) Improving efficiency of groundwater pumping for agriculture: thinking through together. Centre for World Solidarity, Prayas Energy Group, Hyderabad and Pune

Müller M, Rommel J (2013) Technology adoption and energy efficiency in irrigation: first results from a coordination game in Andhra Pradesh. German Association of Agricultural Economists (GEWISOLA), India

Chapter 8
Results

Abstract This chapter provides the main results from the pilot project with respect to technical performance and social aspects. Section 8.1 reports summarised results from an evaluation of the project, with technical data indicating actual observed improvements in power factor and other important parameters, accompanied by a marginal abatement cost analysis. Section 8.2 is more narrative in form, discussing the main observations made by the project team regarding social processes and technical implementation.

Keywords Marginal abatement cost curve · Efficient pumpsets · Evaluation · Power factor · Measurement

8.1 Evaluation Results

8.1.1 Pumpset and DTR Measurement Results

In the following, the central technical findings of the evaluation phase are presented, accompanied by a marginal abatement cost analysis comparing the capacitor's technological potential to other solutions, such as solar water pumpsets, in terms of greenhouse gas (GHG) mitigation and electrical energy savings. For assessing the impact capacitors exert on the grid, the batch-wise method was selected as the most suitable and reliable one (see Sect. 7.3), as it generated the most stable 'before' and 'after' conditions and a higher degree of system equilibrium during the measurement time period as compared to the sequential method. Additionally, relatively constant pumping conditions only needed to be ensured for a 48 h period with the batch-wise method, which was deemed much easier to achieve as opposed to the week-long study period required per DTR for the other methods. For the final analysis, four DTRs were selected from the 16 DTRs connected to the Namiligundupally feeder. Figure 8.1 exemplifies the layout of the pumpsets of one DTR.

© The Author(s) 2016   81
J. Sagebiel et al., *Enhancing Energy Efficiency in Irrigation*,
SpringerBriefs in Environmental Science, DOI 10.1007/978-3-319-22515-9_8

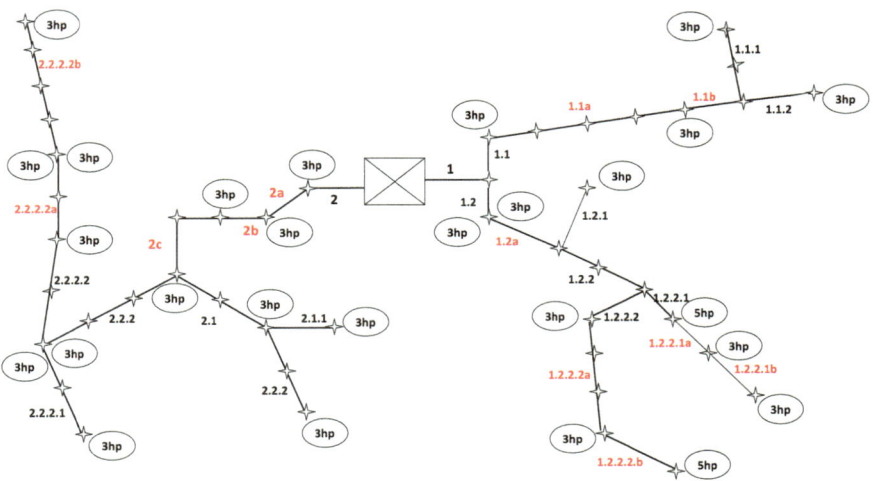

Fig. 8.1 Map of pumpsets connected to the NGP III DTR

Table 8.1 Before and after intervention readings and efficiency parameters from four DTRs

DTR		kW	I	kVA	V	kVAR/kW	kVA/kW	I/kW
KTP I	Before	24.70	167.23	29.36	249.74	0.67	1.21	6.86
	After	22.96	154.49	25.27	250.25	0.46	1.10	6.84
VTM IX	Before	36.90	189.88	47.94	253.61	0.84	1.31	5.13
	After	32.24	152.20	37.46	253.94	0.60	1.17	4.78
NGP II	Before	61.47	322.56	74.58	231.15	0.69	1.22	5.18
	After	70.18	341.20	77.34	227.87	0.47	1.10	4.79
NGP III	Before	55.28	313.97	75.41	243.76	0.90	1.37	5.71
	After	59.67	280.72	69.68	246.44	0.56	1.17	4.79

Table 8.1 displays representative results from four (out of 16) selected DTRs in terms of the readings taken before and after the capacitor intervention. The effects presented here derive from the capacitor model used in phase II of the project.

At first sight, the readings do not seem to deliver a clear picture regarding points of improvement. Voltage levels remained relatively constant, while changes in active power (kW) appeared in both directions. Apart from the NGP II DTR, the measured amount of apparent power, as a sum of real and reactive power, was reduced at the DTR level. This can be mainly attributed to a reduction in reactive power (i.e. increase in power factor) required to operate a given load from pumpsets in the grid. Similarly, no clear pattern can be observed for the system current measured at the DTR level.

However, when efficiency measures are considered, a clear pattern emerges. The amount of reactive power per active power unit (kVAR/kW) was reduced by around 30 % on the DTR level following connection of capacitors to all pumpsets. The results further translate into a reduction of apparent power per power unit (kVA/kW) and also to a reduction of relative current (I/kW), lowering heating of lines and capacitors and the risk of burning of distribution lines and appliances.

Energy conservation achieved due to an amalgamated reduction in line losses (i.e. lower I^2R losses from lower system currents) and reduced reactive power ranged from 6.3 to 24.1 %, with a median of 7.1 %. This equated to approximately 167.3 kWh electrical energy conserved per hp per year. Meanwhile, GHG mitigation achieved at the hp level was approximately 0.2 tonnes of carbon dioxide equivalents (CO_2e) mitigated per hp per year. The monetary benefits to the utility, if it is able to sell the conserved energy to industry, is expected to be in the range of 1500–1600 INR/hp/year. The total conservation achieved due to the pilot installation for one feeder is estimated to be 196,801 kWh/year and 179.1 tonnes CO_2e/year, with a potential financial gross benefit of 1.8 million INR/year for the utility, not considering installation and maintenance costs.

The potential project benefits for the state to be derived from upscaling this intervention are expected to be approximately 1.2 million tonnes CO_2e/year of GHG mitigation, avoided power generation of about 1337 GWh/year and monetary benefits, through sale of power to industry or commercial entities, of roughly 12,206 million INR/year (Tables 8.2 and 8.3). These projections are based on the following system parameters from Central Electricity Authority (2011, Table 9.3):

1. Total motor hp connected per DTR = 1176.5 hp and
2. Annual agricultural energy consumption for the state = 18,825.02 GWh.

8.1.2 Marginal Abatement Cost Analysis

Comparing capacitor-based energy conservation interventions with other plausible demand-side management measures and renewable energy alternatives on a cost–benefit basis is essential for determining its potential for wide-scale impact at the regional, state or national levels.

A marginal abatement cost analysis was conducted for the following demand-side management measures and off-grid renewable energy alternatives:

1. low efficiency pumpset without capacitors (baseline case),
2. low efficiency pumpset with capacitors (case 1),
3. high efficiency pumpset without capacitor (case 2),
4. high efficiency pumpset with capacitor (case 3),
5. low efficiency off-grid solar pumpset (case 4), and
6. high efficiency off-grid solar pumpset (case 5).

The parameters used for the analysis required in order to devise a marginal abatement cost curve for the above interventions are presented in Table 8.4.

Table 8.2 Marginal abatement cost curve analysis for whole Andhra Pradesh

Parameter	Baseline case Low-efficiency pumpset, with-out capacitor	Case 1 Low-efficiency pumpsets, with capacitor	Case 2 High-efficiency pumpset, without capacitor	Case 3 High-efficiency pumpset, with capacitor	Case 4 Low-efficiency solar pumpset	Case 5 High-efficiency solar pumpset	Units
Number of pumpsets	1,592,824	1,592,824	1,592,824	1,592,824	1,592,824	1,592,824	
Horsepower per pumpset	4	4	2.44	2.44	4	2.44	hp
Annual energy consumption	16,907	15,842	10,332	9681	0	0	GWh
Annual GHG emissions	15,386	14,416	9402	8810	0	0	Kilotonnes CO_2e
Total annual operational cost	152,167	142,575	92,991	87,129	0	0	INR/year (in millions)
Total capital cost	19,114	21,294	23,361	24,694	557,488	381,570	INR (in millions)
Energy savings versus baseline case	–	1066	6575	7226	16,907	16,907	GWh/year
GHG savings versus baseline case	–	970	5983	6576	15,386	15,386	Kilotonnes CO2e/year
Operation costs savings versus baseline case	–	9592	59,176	65,038	152,167	152,167	INR/year (in millions)
Capital costs increase versus baseline case	–	2180	4248	5580	538,375	362,456	INR (in millions)
Equipment lifespan	–	10	10	10	10	10	Years
Payback period (simple)	–	0.23[a]	0.39	0.38	3.66	2.51	Years

[a]In this case payback is calculated for incremental cost since pumpset replacement is not required

Table 8.3 Values used for state-level marginal abatement cost curve projection

Parameter	Value	Units
State agricultural energy consumption	18,825[a]	GWh
GHG emission factor[b]	0.91	kg CO_2e/kWh
Energy tariff[c]	9.13	INR/kWh
Energy consumption before intervention	2.91	kWh/hp/h
Average annual operation (at 5 h per day for 6 months per year)	912.50	h
Total pumpset capacity in state	7,093,898	hp
Average pumpset capacity	4	hp
Total pumpsets in state of average capacity	1,592,824	Number of pumpsets
Discount/interest rate for net present value	8 %	

[a]*Source* Central Electricity Authority (2011), Table 9.3
[b]This GHG assessment is based on an Andhra Pradesh-specific GHG Emission Factor of 0.91 kg CO_2e/kWh for 2010–2011 for grid electricity. This accounts for the state-specific fuel mix, India-specific net calorific values (NCVs) for coal as well as the load generation balance data and statistics of inter-state energy transfer provided by Central Electricity Authority (2011)
[c]*Source* http://www.apspdcl.in/ShowProperty/SP_CM_REPO/WhatsNew/Tariff_Order_2013_14, p. 178

Table 8.4 Parametric values used for marginal abatement cost analysis

Parameter	Value	Units
Equipment cost: low-efficiency pumpset	3000	INR/hp
Equipment cost: high-efficiency pumpset	6000	INR/hp
Equipment cost: low-efficiency solar pumpset, unsubsidized	125	INR/hp (in thousands)
Equipment cost: high-efficiency solar pumpset, unsubsidized (assumption: additional efficiency will cost 15,000 INR/hp	140	INR/hp (in thousands)
Solar pumpset subsidy	30 %	
Low-efficiency pumpset hydraulic efficiency	28 %	
High-efficiency pumpset hydraulic efficiency	45 %	
Cost of capacitor + miniature circuit breaker + box	933.33	INR/kVAR
Reactive power	0.37	kVAR/hp
Cost of capacitor + miniature circuit breaker + box per hp	342.22	INR/hp

The consequent marginal GHG abatement cost curve for the alternatives outlined earlier is presented in Fig. 8.2.

The marginal abatement cost analysis presented in Tables 8.2 and 8.5 clearly identifies the basic intervention of installing relatively low-cost capacitors on existing low efficiency pumpsets in the state as being the 'lowest hanging fruit' option. The approximate capital cost for an average 4 hp pumpset would be 13,000 INR, including installation costs and capacitor, while the estimated annual energy, operating cost and GHG emissions savings per year are estimated to be approximately 669 kWh, 6022 INR and 0.61 tonnes CO_2e, respectively (Table 8.5).

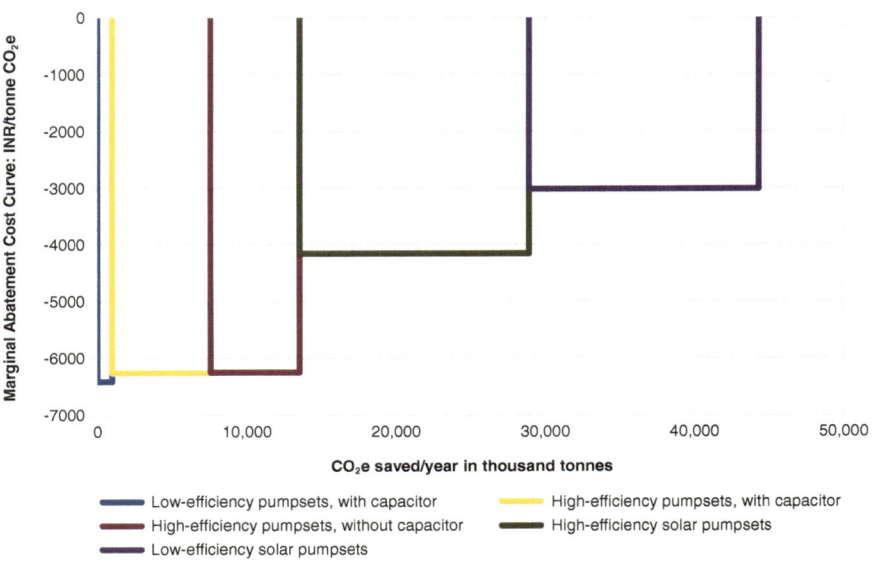

Fig. 8.2 Marginal greenhouse gas abatement cost curve analysis

   While the capacitor addition presents a very lucrative marginal abatement cost opportunity of −6412 INR/tonne CO_2e abated over a simple payback period of 0.23 years, the total mitigation potential through this option for the state—approximately 0.97 million tonnes CO_2e/year and between 1066 and 1337 GWh/year in energy conservation—is modest compared to the other alternatives analysed at a macro level.

   Compared to the simple capacitor addition, the combination of capacitors with a high efficiency pumpset presents an alternative that has equally appealing marginal abatement costs of −6261 INR/tonne CO_2e abated over a simple payback period of 0.38 years, along with a significantly higher total mitigation potential at 6.57 million tonnes CO_2e/year and 7226 GWh/year in energy conservation.

   Finally, high-efficiency solar pumpsets show the least attractive marginal abatement cost value of −4156 INR/tonne CO_2e and a longer payback period of 2.51 years, even after factoring in a 30 % government subsidy. However, they present the maximum potential for GHG mitigation at 15.38 million tonnes CO_2e/year and 16,907 GWh/year in energy conservation.

   In the above marginal abatement cost analysis, it should be noted that the mitigation assessments for capacitors are based on actual field measurements, while the values derived for other demand-side measures (including solar pumpsets) are based on approximate estimates provided by industry representatives associated with the specific technologies (high hydraulic efficiency pumpsets and solar pumpsets). Thus the results have an inherent uncertainty embedded in them, due to the disparity of data sources used for deriving the values shown. In the absence of

Table 8.5 Marginal abatement cost curve analysis for a single pumpset

Parameter	Baseline case Low-efficiency pumpset, without capacitor	Case 1 Low-efficiency pumpset, with capacitor	Case 2 High-efficiency pumpset, without capacitor	Case 3 High-efficiency pumpset, with capacitor	Case 4 Low-efficiency solar pumpset	Case 5 High-efficiency solar pumpset	Units
Total system hp	4	4	2.44	2.44	4	2.44	hp
Annual energy consumption	10,615	9946	6487	6078	0	0	kWh
Annual greenhouse gas emissions	9.66	9.05	5.90	5.53	0.00	0.00	Tonnes CO_2e
Total annual operational cost	96	90	58	55	0	0	INR/year (in thousands)
Total capital cost	12	13	15	16	350	240	INR (in thousands)
Energy savings versus baseline case	–	669	4128	4537	10,615	10,615	kWh/year
Greenhouse gas savings versus baseline case	–	0.61	3.76	4.13	9.66	9.66	Tonnes CO_2e/year
Operation cost savings versus baseline case	–	6	37	41	96	96	INR/year (in thousands)
Capital cost increase versus baseline case	–	1	3	4	338	228	INR (in thousands)
Equipment lifespan	–	10	10	10	10	10	Years
Payback period (simple)	–	0.23[a]	0.39	0.38	3.66	2.51	Years

[a]In this case payback is calculated for incremental cost since pumpset replacement is not required

actually conducting comparable pilot studies with the other possible technological interventions, we believe that the method explained above is the best one available for conducting such an analysis. Consequently, it is recommended that the results be seen in this context as indicators of the relative attractiveness of the various agricultural energy-related GHG mitigation options, rather than as sources of exact marginal abatement cost values for the options described. They are designed to serve as decision-support inputs for macro-level policy formulations and would require additional pilot testing to yield much more accurate marginal abatement cost values of a quality similar to those for the capacitor intervention.

8.2 Observations from the Field

8.2.1 Social Implementation

The social aspect of the pilot project was expected to play a crucial role for the project's success. Hence, an emphasis was put on farmer mobilisation. For this, a local social worker and consultant was employed to oversee the project's overall preparation and implementation. Further, the local NGO, SEWS, was given responsibility for sound social implementation. This strategy was intended to assure that the technical team would not face significant social problems during implementation and evaluation. However, after the technical problem of the malfunctioning capacitors occurred, the cooperativeness of farmers in the Sangula area declined sharply—so much so that the social intervention team could not manage to restore trust. Yet, the farmers in Namiligundupally remained cooperative, and the overall mood was still positive. The reason for this seemed obvious: farmers in Namiligundupally have been participating for more than ten years in the local watershed program, working together with SEWS, and hence had more trust in the project.

This observation sheds light on the necessity to integrate local actors into such projects, as intense and long-term relationships with stakeholders and participants are required to build resistance against unexpected and adverse events. Furthermore, the observation demonstrates the interaction between social and technical factors: such social problems as occurred in Sangula would very likely not have arisen if the technical problems had been absent and, conversely, with proper social implementation, as in Namiligundupally, technical problems can be absorbed. All things considered, the social implementation of the project could have been improved. Although the project team put high priority on this, it did not make the project resistant to unexpected exogenous shocks such as, in this case, the burning of capacitors.

8.2.2 Technical Implementation

Technical implementation of the project greatly depended on farmer cooperativeness. It turned out that the awareness meetings that had been conducted before installation were crucial. Many farmers asked critical questions and were, initially, rather reluctant to adapt to the technology. With open discussions, however, ambiguities were cleared up and reservations overcome. Hence, the support from CESS and SEWS turned out to be a key driver for an unproblematic implementation process, as SCTI did not face major difficulties from the farmers' side when installing the capacitors.

During installation, however, several unexpected difficulties did occur. Unfavourable positioning of the starter boxes, for example, hindered sound installation so that, in some cases, it was not possible to install a capacitor at all. Also, capacitor failure during the process was an issue, with farmers becoming sceptical and sometimes not permitting their re-installation. Most of these problems could be solved directly or with help from the social team. During the first months of installation, a decent level of learning progress took place. Installing the first capacitors took a long time—with only about two per person and day being installed—and some safety precautions were not followed by the electricians. When this problem was detected, another round of training took place. Consequently, overall performance increased and the number of installed capacitors per day was raised to between three and five capacitors per person. Thus, by the time of implementation phase II, the staff had gained sufficient experience to complete the installation of about 400 capacitors within one month.

Reference

Central Electricity Authority (2011) All India electricity statistics 2011—general review. New Delhi

Chapter 9
Upscaling Potential

Abstract Upscaling of small projects is an important step for reaching widespread application and properly capturing estimated effects in terms of electrical energy savings and reduced carbon dioxide emissions. In this chapter, we discuss alternatives for upscaling of this project in India, including regional upscaling (i.e., extend the project to other areas), technical upscaling (i.e., increase the number of technologies used), and political upscaling (i.e., give policy recommendations for regulation reforms that would enable more projects like this). Further, we discuss different business models that can be applied to make the project financially sustainable.

Keywords Regional upscaling · Technical upscaling · Political upscaling · Business models · Regulation reforms in India

9.1 Regional Upscaling

Different plans to upscale the project within and also to other regions were proposed to the involved stakeholders. First, discussions were initiated with CESS to install capacitors across the whole CESS region, which would cover about 60,000 agricultural connections. Further discussions with officials from governmental departments, the regulatory commission, and distribution companies indicated that the idea to initiate a larger governmental program with capacitors is a realistic option for the future. Another option for regional upscaling models is to incorporate the project into existing agricultural or rural development projects. Solid demonstration of the effects of capacitors in further studies may help to convince stakeholders of the financial viability of this option.

© The Author(s) 2016   91
J. Sagebiel et al., *Enhancing Energy Efficiency in Irrigation*,
SpringerBriefs in Environmental Science, DOI 10.1007/978-3-319-22515-9_9

9.2 Technical Upscaling

Once having established a firm social structure, further technologies can then be easily implemented. When the farmers are ready to invest in demand-side management measures, this can be done without a need for great financial support. Once the capacitor solution has been adopted, the following options can be readily implemented:

- using heavy-duty capacitors for longer life and better reliability;
- individually adjusting capacitor rating to improve PF compensation;
- placing miniature circuit breakers at each terminal, running from capacitor to supply/motor terminals;
- installing dry-run preventers to switch motors off when no or insufficient water is available;
- installing energy meters for each pumpset, for better identification of power consumption and fluctuations;
- upgrading pumpsets to more efficient ones, including BEE-rated ones;
- improving repair service by, for example, training local repair shop personnel about better rewinding techniques; and
- installing solar photovoltaic systems to backup grid supply.

9.3 Business Models for Upscaling

The field research phase, the social and technical evaluation data, as well as the experience and learning gained from the project have enabled the development of a business plan. The transaction costs, including training, technology installation and governance unit formation, can be calculated to derive the profitability of the overall project. The business plan can then be utilised by Energy Service Companies and other contractors to conduct energy-efficiency improvement projects. Counterparties to such contractors can be electricity utilities as well as the government agencies that ultimately pay for the subsidised electrical energy provided for irrigation. A share of the saved energy expenses can then be used as revenues to finance the contractor.

9.4 Political Upscaling

Policy briefs and consultations can enable direct communication with responsible government units, including energy departments and regulatory commissions, to inform them about design of more effective policies for demand-side management measure implementation. Such policies can include cooperation with contractors for grassroots implementation, due to their expertise in required technologies and entrepreneurial skills. The regulatory framework needs to provide regulations and credible commitment for facilitating efficiency-enhancing contracts between contractors and utilities that can be profitable. It remains to be seen, however, whether contractors or other organisations are able to create the institutional arrangements necessary at the DTR level to facilitate successful coordination of technology adoption.

Chapter 10
Conclusions and Outlook

Abstract The final chapter of this SpringerBrief summarizes the project details and outcomes. It briefly discusses further steps that farmers, utilities and governments could undertake to increase energy efficiency in agriculture, thereby reducing electrical energy consumption and CO_2 emissions.

Keyword Andhra Pradesh · Energy efficiency in agriculture · Capacitor · Rural development · Groundwater irrigation

This Springer Brief has been divided into two parts. Part I gave general background information on the power sector in India, issues inherent to agricultural electrical energy supply, options that could be expected to mend or at least improve the current situation, and an overview of ongoing and recent projects. The purpose of this part was to introduce readers to the topic, make them aware of the complexity of the problem and demonstrate why, until now, no "way out" of the vicious circle of agricultural electrical energy supply has been found in India. Furthermore, the authors also aimed at providing enough background information so that a general readership can understand the issues raised in Part II.

Part II summarized the implementation of a pilot project for improving energy efficiency in Indian agriculture. The research phase started in 2008 and a first project design was developed in 2010. From 2011 to 2013, a project consortium from different fields, including social scientists and economists as well as technical experts, implemented and evaluated the project. The main idea of the project was to bring an inexpensive yet effective technology to the (agricultural) field, observe how well it operates, how it gets accepted by farmers and under what conditions it fails. The technology of choice was the application of shunt capacitors directly installed at each load, meaning the motors of water pumpsets used for groundwater irrigation. To complement the technical approach, social interventions took place through which farmers were motivated to form groups at their respective distribution transformers to better manage the technology adoption and issues on DTR level.

J. Sagebiel et al., *Enhancing Energy Efficiency in Irrigation*,
SpringerBriefs in Environmental Science, DOI 10.1007/978-3-319-22515-9_10

In this part, the pilot project was introduced through a detailed description of project partners, stakeholders and the location chosen for implementation. Then, the aims and rationale of the project were explained, as were the reasons why capacitors were chosen: first, they would be affordable to the farmers; second, they could be installed rather easily with help from local mechanics; and, third, they would benefit both the local utility and the farmers. The utility would benefit from reduced line losses, increased power factor at substation level, less energy consumption by agricultural users and reduced repair costs for burned-out distribution transformers. Farmers would be able to take advantage of better voltage levels, leading to a reduced number of distribution and motor burnouts and, in some cases, increased water flows.

To better understand how the project was implemented, each step was explained in detail. The first, preparatory, phase had been concerned with selection of partners, technology and location. Two workshops had been conducted where experts discussed different strategies for implementation. The choice of feeders was the result of one month of research, including data collection at all feasible feeders and, eventually, a ranking was produced to select the most appropriate feeder. The choice of capacitors followed a similar procedure. Different capacitor manufactures were contacted and, based on certain criteria, one company was chosen. As already mentioned, the implementation phase included social and technical interventions, during which the capacitors were installed and farmers groups established. Farmers had been continuously briefed on the project's status and made aware of the benefits of capacitors. After installation, farmer committees were founded. Farmers received training, conducted regular meetings, interacted with the utility and recorded important variables from the field, including number of motor burnouts. As there had been problems with the first batch of capacitors, and it turned out that the capacitors could not withstand the heavy voltage imbalances, all capacitors were replaced with a better model and protected with additional safety and monitoring equipment.

The performance of the new capacitors was then evaluated via an extensive measurement process at selected DTRs, with results indicating an increased power factor of about 16 %. A marginal abatement cost analysis subsequently revealed that capacitors are very likely the most cost-efficient solution for saving on electrical energy and GHG emissions, but the total potential is limited, as overall savings are not as great as with other technologies.

Finally, the potential for upscaling was discussed, aiming towards extending the project's findings to the whole working area of the Co-operative Electric Supply Society Sircilla, Ltd. and again evaluate them. Furthermore, it was suggested that other more-advanced technologies could also be tested and evaluated. At bottom, the long-term aim should be to find a holistic approach that can have lasting effects on energy quality and consumption as well as on CO_2 emissions yet withstand expected social, political and technical problems that may arise.

Appendix I
Technical Parameters

Voltage (V) Voltage is the force/pressure in the circuit which drives the Electrons. It is measured in Volts (V). The voltage that we are dealing with for pumpsets under Namiligundapally and Sanugula feeders are 3 phase 3 wire 440 V system.

$$V = I \times Z$$

where V = Voltage, I = Current, Z = Impedance

Current Current is defined as the rate of flow of charge. It is measured in Amperes (A). The amount of current drawn from the source will depend on the connected Load.

Power Factor Power Factor is defined as the ratio of Real Power to Apparent Power. It is also defined as the cosine of angle between Voltage and Current. Generally

Power Factor = Real Power (Watts)/Apparent Power (VA)
Power Factor = Cos (angle between V and I)

Power Power is defined as the amount of energy used or converted per unit time. It is measured in Joules/sec (J/s) or Watts (W).

$$\text{Real Power} = V \times I \times \cos\theta$$

$$\text{Reactive Power (kVAR)} = V \times I \times \sin\theta$$

$$\text{Apparent Power} = V \times I$$

where V = Voltage, I = Current, θ = Angle between V and I

© The Author(s) 2016
J. Sagebiel et al., *Enhancing Energy Efficiency in Irrigation*,
SpringerBriefs in Environmental Science, DOI 10.1007/978-3-319-22515-9

Energy The capacity to do work is called Energy.

Impedance (Z) Impedance is the AC-circuit equivalent of Resistance (used when addressing DC circuits). It is defined as the magnitude of resistance to current flow in a circuit when a voltage is applied. It is measured in Ohms (Ω).

$$Z = V/I$$

Impedance is markedly different from resistance as it includes the conventional 'resistive' aspect of DC circuits and adds to two additional effects which resist current flow—resistance from self-induced magnetic fields generated by voltages in the circuit (called inductance), and the resistance from electrostatic charges stored by voltages between conductors (known as capacitance). The sum of these is called 'reactance'. Impedance then is the vector sum of these two quantities—resistance and reactance.

The primary energy conservation and GHG emissions reduction impact of reduced reactive power in distribution systems materializes in the form of reduced energy generation required to supply a given amount of 'real' power. Reduced reactive power reduces the amount of current required to be carried in the system to deliver a given quantity of active power (kW). This reduced current reduces the line losses (due to resistance of the conductor) from the system which thereby conserves the consumption of fuel at a generation station. The relationship is shown below:

Power Generation (kW) = Resistive Line Losses (kW)

+ Delivered Active Power (kW)

$$R = r \times L/A$$

where,

L = Length of Conductor
A = Area of cross section of Conductor
r = Resistivity

$$\text{Resistive Line Losses (W)} = 3 \times R \times L \times I^2$$

where,

I = current per conductor or phase (A)
R = resistance per conductor or phase (Ω/km)
L = length of each segment (km)

$$\text{Conservation \%} = \frac{I2Rcons._{DTR}}{(kW_{DTR} + I2R_{DTR}) \times Hours_{annual\ energy\ supply}}$$

where,

$I2R\,cons._{DTR} = DTR\,level\,I2R\,reduction(kW) \times Hours_{annual\,energy\,supply}$

kW_{DTR} = active power generation required at plant to power DTR (excludes reactive energy)

$I2R_{DTR}$ = total losses related power generation required at plant to support transmission and distribution losses

Note: $Hours_{annual\,energy\,supply,i}$ assumed to be equal to 912.5 h for all pumpsets

$$\text{Voltage Drop} = \sqrt{3} \times L \times R \times I$$

where,

L = length of wire (km)
R = conductive resistance (Ω/km)—per phase
I = current (amps)—per phase

Appendix II
Technical Questionnaire

Pump set level data

	VARIABLE	DATA/INFORMATION	Remarks		
1	Investigator Name:				
2	Date:				
3	11 KV Feeder Name:				
4	Village Name:				
5	DTR Number				
6	Pump set ID:				
7	Owner name				
8	Well Type:	Shallow well / Borewell			
9	Pump Type	Centrifugal / Submersible			
10	Pump Rating	3 / 5 / 7.5 HP			
11	Control panel	Available / Not Available			
12	Capacitor connected :	1 / 2 / 3 KVAR			
13	Earthing :	OK / Not OK			
		Pump Earthed /Capacitor Earthed / Pump and Capacitor Both Earthed			
14	Over all Condition of Pump Installation:	Good / Not Good			
15	Maintenace of Pump Set:	Good / Not Good			
16	Surveyor's opinion:	Good / Not Good			
17	Time: AM/PM	Pump Level Reading - Before Capacitor Installation			
		R	Y	B	
	Voltage				
	I (Current)				
	KW				
	KVA				
	KVAR				
	PF				

© The Author(s) 2016
J. Sagebiel et al., *Enhancing Energy Efficiency in Irrigation*,
SpringerBriefs in Environmental Science, DOI 10.1007/978-3-319-22515-9

18	Time:	AM/PM	Pump Level Reading - After Capacitor Installation		
			R	Y	B
	Voltage				
	I (Current)				
	KW				
	KVA				
	KVAR				
	PF				
19	Time:	AM/PM	Pump Level Reading - After Capacitor Installation		
			R	Y	B
	Voltage				
	I (Current)				
	KW				
	KVA				
	KVAR				
	PF				

20	Time:	AM/PM	Pump Level Reading - After Capacitor Installation		
			R	Y	B
	Voltage				
	I (Current)				
	KW				
	KVA				
	KVAR				
	PF				
21	Water Output Measurement through 'L' gauge		Possible / Not Possible		
	If not possible, reason:				

Time		cms	Liter / sec (Calculated)
	AM / PM		
	AM / PM		
	AM / PM		
	AM / PM		

Notes:

DTR level data

	VARIABLE	DATA/INFORMATION	Remarks		
1	Investigator Name:				
2	Lineman Name:				
3	Date:				
4	11 KV Feeder Name:				
5	Village :				
6	DTR Number:				
7	Pump set ID:				
	DTR Level Reading - Before Capacitor Installation				
8	Time: AM/PM	R		Y	B
	Voltage				
	I (Current)				
	KW				
	KVA				
	KVAR				
	PF				
	DTR Level Reading - After Capacitor Installation				
9	Time: AM/PM	R		Y	B
	Voltage				
	I (Current)				
	KW				
	KVA				
	KVAR				
	PF				
10	Time: AM/PM	R		Y	B
	Voltage				
	I (Current)				
	KW				
	KVA				
	KVAR				
	PF				
11	Time: AM/PM	R		Y	B
	Voltage				
	I (Current)				
	KW				
	KVA				
	KVAR				
	PF				
12	Time: AM/PM	R		Y	B
	Voltage				
	I (Current)				
	KW				
	KVA				
	KVAR				
	PF				

Notes:

Appendix III
DTRC Constitution and Minutes of Meetings

Minutes of the Meetings
Log Book

DTR Committee name:

Village: Mandal: District:

President:
Secretary:
Treasurer:

Committee size: Year:

© The Author(s) 2016
J. Sagebiel et al., *Enhancing Energy Efficiency in Irrigation*,
SpringerBriefs in Environmental Science, DOI 10.1007/978-3-319-22515-9

A.   SHORTLIST OF MEETINGS

No.	Date	Venue	Convened by/on	People attended	Page	Signature Secretary
1.						
2.						
3.						
4.						
5.						
6.						
7.						
8.						
9.						
10.						
11.						
12.						
13.						
14.						
15.						
16.						
17.						

B. Constitution Distribution Transformer Committee

§ 1 Name of the DTRC

The committee is named after the DTR name:_____

§ 2 Purpose

The purpose of the committee is to improve and maintain the electricity supply quality in order to reduce costs. The core aim of the DTRC is to work towards a contract with CESS. The DTRC commit itself to improve power factor of DTRs and reduce DTR burn outs. In turn CESS will grant a discount of the monthly connection fee of 75 % to all active members of the DTRC. In the long run all illegal connections shall be regularized and motors shall not be overrated. Apart from this, side activities shall be considered. These may include training on how to maintain a capacitor, how to maintain a motor, where to rewind. Also, activities not related to electricity can be included on a voluntary basis like improved irrigation techniques, savings groups, collective marketing and input buying, extension etc.

§ 3 Group Formation (General Body)

To participate in the DTRC, all farmers connected to the respective DTR (20–30 farmers) are eligible. All members shall have at least one water pump and are dependent on groundwater Irrigation. Farmers, who are connected to other DTRs shall not participate in this DTRC. The members of the DTRC are collectively responsible for maintenance of DTRs. The contract with CESS shall be fulfilled by the farmers, and all farmers should be aware of their rights and duties before group formation. The DTRC shall select representatives (Managing Body) and these representatives shall fulfill certain roles.

§ 4 Managing Body

The managing committee shall consist of the President, Secretary and Treasurer. Managing Body shall have power to appeals and raise funds and fulfill and formalities incumbent upon it.

§ 4.1. President He/She shall be in overall charge of the committee and the general body meetings. All the policies and programs shall be formulated and implemented only through him/her. He/She shall operate a bank account jointly with the Treasurer. The President of a committee has four main elements to his/her remit as follows:

- Assisting with the managerial direction of the DTRC
- Planning and running meetings
- Acting as spokesperson/figurehead
- Communicate with the Feeder Committee

§ 4.2. Secretary He/She shall call for all general body meetings as and when deemed necessary and the General body meetings and the Special body meeting as per the rules with the previous approval of the president and maintain the minutes of meeting (MoM) book and record of all the proceedings of the meetings. He/She shall be the correspondent of the committee and shall be in-charge of the office with all the record of the DTRC. He/She shall be the custodian of all articles and belonging both movable and immovable of the committee. The Secretary's main responsibilities are:

- Supporting the administration of the DTRC

- Facilitating and supporting DTRC meetings
- Together with the President, correspond with the Feeder Committee

§ 4.3. Treasurer The Treasurer has the day-to-day responsibility for looking after DTRC's money. However the DTRC as a whole is responsible for deciding how funds will be raised and spent. His/her job is to keep accounts, collect ingoing and outgoing receipts and report to the committee. He/She shall operate bank account jointly with the President. The Treasurer also has three main areas of responsibility:

- Keeping an overview of the finances of the DTRC
- Reporting into DTRC meetings
- Making sure the DTRC has the right financial policies and procedures in place

§ 5 Technical Assistant

The technical assistant should be selected based on his (electro) technical skills. He should be aware of general principals of motors, pump sets, capacitors and DTRs. Further he should participate in regular technical training sessions held at the Pilot Project Committee. He should consult his fellow farmers in technical problems and carry out small repair works. He shall get additional payment based on his works done to other farmers. His main tasks are:

- Overviewing the technical health of the DTR and LT grid
- Assisting his fellow farmers with technical problems
- Participating in regular trainings
- Maintaining technical record (log book)
- Communicate with Feeder Committee on technical topics

§ 6 Bank Account

The DTRC will open a separate bank account for all ingoing and outgoing payments from and for the members. The bank account shall be overviewed and maintained by the Treasurer and the Secretary or the Treasurer and the President.

§ 7 Roles of DTRC Members

Committee members are responsible for stimulating and instigating discussion with other farmers regarding future activities. Each DTRC member is responsible to maintain his capacitor and to maintain the DTR as per the instructions by the technical assistant. Each member shall spend at least two hours per week with voluntary work to maintain the DTR. He shall follow the instructions by the technical assistant. Each member shall actively participate in the regular DTRC meetings and provide necessary data to the MoM as per instructions of the secretary. His/Her roles are:

- Actively participate in DTRC meetings
- Provide data to the Minutes of Meeting

- Maintain his capacitor as instructed by technical assistant
- Offer some of his work time to maintain the DTR

§ 8 Meetings

Meetings are the core ground of the DTRC. The meetings should include train-ing sessions, social *and* technical presentations, and general discussion. All meet-ings shall be documented in the Minutes of Meeting (MoM). The DTRC shall be equipped with a MoM book.

§ 8.1. Frequency of Meetings Regular meetings shall be held monthly once. Occasionally, it may be necessary to call an extraordinary meeting of the DTRC. It is important to note that the purpose of an extraordinary meeting must be clearly stated when the request for it is made. The agenda for the meeting should only contain papers directly relevant to the issue(s) under discussion. The meeting is not asked to approve minutes, deal with matters arising, nor will discussion of other issues be allowed

§ 8.2. Duration The duration of the meeting should be one hour. It can be extended if necessary.

§ 8.3. Timings The meetings should take place in the morning.

§ 8.4. Venue A common meeting place can be suggested by DTRC.

§ 8.5. Topics discussed The topics discussed shall be based on the MoM. The agenda will be prepared by NGO staff and president. Each meeting shall include:

- Training session (Technical staff)
- Social Presentation (NGO staff)
- Technical Presentation (Technical staff)
- Discussion
- Documentation

§ 8.6. Minutes of Meeting DTRC Secretary is responsible for distributing hard copies of the MoM to committee members.

§ 8.7. Quorum The quorum of the meeting shall be 1/3rd of the total membership of the committee.

§ 8.8. Collection of fees The treasurer shall collect the fees from all members at the beginning of the meeting.

§ 9 Fees

An amount of Rs. 50/- per month from each farmer is suggested but can be increased or reduced by the farmer. The fees will be maintained by the treasurer and used for general DTRC expenses, common repair work and savings. Additionally a yearly fee of Rs 100/- for maintenance of the DTRC should be paid by the members.

- Yearly Membership fees Rs. 100/-
- Monthly saving Rs. 50/-

§ 10 Election

Voting shall be conducted by show of hands or secret ballot. Managing Body member shall undertake the role for a minimum of one year, with each member to have the option of re-nominating following completion of the term. To retain, appoint, promote, and dismiss any member for managing and functioning of the DTRC.

§ 11 Support

Support to the DTRCs and farmers will be available through ITI guys for technical issues and SEWS for social related issues.

§ 12 Illegal Connections

The DTRC shall aim to reduce the number of illegal connections. The DTRC members shall openly discuss the issue of illegal connection and shall find ways to regularize them. This process shall take no longer than one year after formation of the DTRC.

§ 13 Relationship with Feeder Committee:

The Feeder Committee is the head organization of the DTRC. The DTRC shall provide some financial contribution to the Feeder Committee. The Managing Body shall participate in regular Feeder Committee meetings and contribute to the discussion. The Feeder Committee shall be also the body for complaints that cannot be solved in the DTRC.

§ 14 Relationship with the Pilot Project Committee

The Pilot Project Committee (PPC) is the head organization of the Feeder Committee. The DTRC is related to the PPC through the Feeder Committee. DTRC managing body shall participate in any PPC meetings and the DTRC shall obey the decisions taken by the PPC. The PPC shall be also the body for complaints that can neither be solved in the DTRC nor in the Feeder Committee.

1. PROTOCOL OF THE 1ST MEETING

Date: Time: Venue:

People attended: People missing: Necessary quorum:

1.1. Social Presentation   Held by:
Role:

_____

_____

_____

_____

_____

_____

_____

_____

_____

_____

1.2. Technical Presentation   Held by:
Role:

_____

_____

_____

_____

_____

_____

_____

_____

_____

_____

1.3. Farmers discussions (open problems)

_____

_____

_____

_____

_____

_____

_____

Problem specification	Occurred since	Involved people	Proposed solution	Recorded in Log Book?
				Yes / No
				Yes / No
				Yes / No
				Yes / No
				Yes / No
				Yes / No
				Yes / No

1.4. Actions taken & planned

_____

_____

_____

_____

_____

_____

Action	Responsible (name)	Completed on
e.g.: Capacitor acquisition & Costs		

1.5. Trainings conducted

_____

_____

_____

_____

_____

_____

_____

_____

_____

_____

1.6. Log Book

S.no	Complaint of Farmer	Name of farmer affected	Date on which problem has been Reported/Occurred	Time of problem Occurred	Name of electrician attending the problem/complaint

Log Book (ctd.)

Pump No	Problem code	Solution	Problem solved Y/N	Date of problem solved	Comments

Signature President:_____

1.7. <u>Attendance signatures:</u>

President: _____

Secretary: _____

Treasurer: _____

Farmers:

_____   _____

_____   _____

_____   _____

_____   _____

_____   _____

_____   _____

_____   _____

_____   _____

_____   _____

_____   _____

1.8. <u>Technical Observations (for this month)</u>

Variable	No	Time of measurement
No of Motor burnouts		
No of Capacitor Failures		
No of DTR burnouts		
PF at DTR		
Voltage at DTR		
Current at DTR		
Further Observations		

Appendix IV
Letter of Agreement to Join DTRC

Letter of Agreement of the Farmer to become member of Distribution Transformer Committee

Agreement taken by : DTR Committee
VILLAGE _____DTR NO _____
MANDAL _____
Agreement given by : Mr. / Mrs._____
VILLAGE _____ DTR NO _____
MANDAL _____

I, a farmer of the above mentioned village, state that there is a need forDistribution Transformer Committees (D⁻RC) in our village. To successfully implement the Programme in our village, I agree to observe the rules and conditions as mentioned below:

1. I will willingly participate in all types of meetings held in the DTRC and contribute to the discussions.
2. I will contribute my energies and time to making the DTRC a success.
3. I will contribute at least two hours labour per week as shramdan (voluntary labour) to the DTRC.
4. I will enforce myself to maintain my capacitor as per the instructions from the DTRC.
5. I will keep away from all types of disputes, at personal or community level, based on caste, religion, class, politics or difference of opinion, which may affect the community. If any such dispute occurs in the village, I will provide help in settling it at the village level itself through democratic processes.
6. As I understand that electricity is a scarce resource, I am agreeable to its equitable distribution on principles decided by the DTRC. I will not put illegal connections and I will not increase the capacity of my motor without reporting to the DTRC
7. We shall be in agreement with all decisions taken for the successful implementation of the project taken by the DTRC, the Feeder Committee or the Pilot Project Committee.
8. I am in full agreement with the above rules and conditions. I understand that these are binding on me and my family members. If any of these rules are violated by me or my family members, I would be fully responsible and any decision taken on this violation by the DTRC the Feeder Committee or the Pilot Project Committee would be acceptable to me.

I am signing this agreement letter, based on full understanding and on my own choice, on
_____, date / /20 , in the presence of witnesses.

Signature of Farmer

Signature DTRC President

Name of Witnesses Signature
1.
2.

J. Sagebiel et al., *Enhancing Energy Efficiency in Irrigation*,
SpringerBriefs in Environmental Science, DOI 10.1007/978-3-319-22515-9